Disaster Diplomacy

When an earthquake hits a war zone or cyclone aid is flown in by an enemy, many ask: Can catastrophe bring peace? Disaster prevention and mitigation provide similar questions. Could setting up a flood-warning system bring enemy countries together? Could a regional earthquake building code set the groundwork for wider regional cooperation?

This book examines how and why disaster-related activities do and do not create peace and reduce conflict. Disaster-related activities refer to actions before a disaster such as prevention and mitigation along with actions after a disaster such as emergency response, humanitarian relief, and reconstruction. This volume investigates disaster diplomacy case studies from around the world, in a variety of political and disaster circumstances, from earthquakes in Greece and Turkey affecting these neighbours' bilateral relations to volcanoes and typhoons influencing intra-state conflict in the Philippines. Dictatorships are amongst the case studies, such as Cuba and Burma, along with democracies such as the USA and India. No evidence is found to suggest that disaster diplomacy is a prominent factor in conflict resolution. Instead, disaster-related activities often influence peace processes in the short term – over weeks and months – provided that a non-disaster-related basis already existed for the reconciliation. That could be secret negotiations between the warring parties or strong trade or cultural links. Over the long term, disaster-related influences disappear, succumbing to factors such as a leadership change, the usual patterns of political enmity, or belief that a historical grievance should take precedence over disaster-related bonds.

This is the first book on disaster diplomacy. Disaster-politics interactions have been studied for decades, but usually from a specific political framing, covering a specific geographical area, or from a specific disaster framing. As well, plenty of quantitative work has been completed, yet the data limitations are rarely admitted openly or thoroughly analysed. Few publications bring together the topics of disasters and politics in terms of a disaster-diplomacy framework, yielding a grounded, qualitative, scientific point of view on the topic.

Ilan Kelman is a Senior Research Fellow at the Center for International Climate and Environmental Research – Oslo (CICERO).

Disaster Diplomacy
How disasters affect peace and conflict

Ilan Kelman

LONDON AND NEW YORK

First published 2012
by Routledge
2 Park Square, Milton Park, Abingdon, Oxon OX14 4RN

Simultaneously published in the USA and Canada
by Routledge
711 Third Avenue, New York, NY 10017

Routledge is an imprint of the Taylor & Francis Group, an informa business

© 2012 Ilan Kelman

The right of Ilan Kelman to be identified as author of this work has been asserted by him in accordance with sections 77 and 78 of the Copyright, Designs and Patents Act 1988.

All rights reserved. No part of this book may be reprinted or reproduced or utilised in any form or by any electronic, mechanical, or other means, now known or hereafter invented, including photocopying and recording, or in any information storage or retrieval system, without permission in writing from the publishers.

Trademark notice: Product or corporate names may be trademarks or registered trademarks, and are used only for identification and explanation without intent to infringe.

British Library Cataloguing in Publication Data
A catalogue record for this book is available from the British Library

Library of Congress Cataloging-in-Publication Data
Kelman, Ilan.
Disaster diplomacy / Ilan Kelman.
p. cm.
Includes bibliographical references and index.
1. Disaster relief. 2. Disasters--Prevention. 3. Humanitarian assistance.
4. Conflict management. 5. Political violence--Prevention. I. Title.
HV553.K425 2011
363.34'56--dc22
2011005711

ISBN: 978-0-415-67993-0 (hbk)
ISBN: 978-0-203-80621-0 (ebk)

Typeset in Times
by Integra Software Services Pvt. Ltd, Pondicherry, India

Printed and bound in Great Britain by
CPI Antony Rowe, Chippenham, Wiltshire

Contents

List of tables	vii
1 The origins of disaster diplomacy	1
1.1 Introduction 1	
1.2 A brief history of disaster diplomacy 4	
2 Moving forward with disaster diplomacy	7
2.1 What this volume offers 7	
2.2 What this volume does not offer 8	
2.3 The structure of this volume 9	
3 Hypotheses and research questions	11
3.1 Definitions of disaster diplomacy 11	
3.2 Hypothesis: catalysis, not creation 13	
3.3 Questions for disaster diplomacy 15	
4 Empirical evidence: Case studies	18
4.1 Organising case studies 18	
4.2 Iran–USA from 1990 onwards 20	
4.3 The Philippines from 1990 onwards 24	
4.4 Southern Africa 1991–93 26	
4.5 North Korea from 1995 onwards 27	
4.6 Cuba–USA from 1998 onwards 30	
4.7 Greece–Turkey from 1999 onwards 32	
4.8 Eritrea–Ethiopia 2000–02 34	
4.9 India–Pakistan in 2001 and 2005 36	
4.10 26 December 2004 tsunamis: Sri Lanka and Aceh 40	
4.11 26 December 2004 tsunamis: other locations 44	
4.12 Hurricane Katrina in 2005 48	
4.13 Two May 2008 disasters 52	
4.14 Island evacuation due to sea-level rise 55	
4.15 Disaster-casualty identification 60	

4.16 International vaccination programmes 63
4.17 Summing up the case studies 66

5 Analyses and typologies for disaster diplomacy 69

5.1 Quantitative analyses 69
5.2 Qualitative typologies 77
5.3 No predictive model 98
5.4 Summarising the typologies 100

6 Explaining disaster diplomacy's successes 102

6.1 Success pathways 102
6.2 Further success: tit-for-tat 106
6.3 Further success: mirror disaster diplomacy 108

7 Explaining disaster diplomacy's failures 110

7.1 Failure pathways 110
7.2 Further failure: inverse disaster diplomacy 113
7.3 Further failure: disaster-related activities exacerbating conflict 114

8 Spin-offs 119

8.1 Environmental diplomacy 119
8.2 Para-diplomacy and beyond 122

9 Limitations 127

9.1 Ethics 127
9.2 Confounding factors 131
9.3 Bias 133

10 Principal lessons for application 135

10.1 Be ready for assistance offers from enemies 136
10.2 All diplomacy tracks can be useful 137
10.3 Disaster diplomacy operates at many levels 139
10.4 Lessons should be implemented, not forgotten 140

11 Filling in the gaps 142

11.1 Can the limitations be overcome? 142
11.2 Why further study disaster diplomacy? 144
11.3 Main gaps to be overcome 146

12 The future of disaster diplomacy 147

References 150
Index 166

List of Tables

4.1	Disaster-diplomacy case studies in this chapter (in alphabetical order by case-study name)	19
4.2	Examples of Hurricane Katrina aid offered by states in conflict with the USA	51
4.3	Summary of disaster-diplomacy case studies and research questions	67
5.1	The complex-adaptive-systems approach applied to three disaster-diplomacy case studies	80
5.2	Propinquity evident in disaster diplomacy case studies	86
5.3	Aid relationships evident in disaster-diplomacy case studies	87
5.4	Diplomacy tracks followed in disaster-diplomacy case studies	89
5.5	Disaster-diplomacy examples for type of state involvement	92
5.6	Active and passive disaster diplomacy	96
5.7	Summary of disaster-diplomacy typologies and categories	97
6.1	Pathways promoting disaster diplomacy	103
7.1	Pathways inhibiting disaster diplomacy	111

1 The origins of disaster diplomacy

1.1 Introduction

The world spends over $1.4 trillion a year in recorded military expenditures (SIPRI, 2009), but less than 1.5 per cent of that on net official humanitarian assistance (GHA, 2009; OECD, 2008). That financial support for war produces results. The UCDP/PRIO Armed Conflict Dataset (Gleditsch *et al.*, 2002; Harbom and Wallensteen, 2009) listed 37 armed conflicts around the world for 2008, covering approximately one fifth of the world's countries. Battle deaths in armed conflicts totalled about 10 million from 1946 to 2002, although battle deaths account for only a small fraction of war-related deaths (Lacina and Gleditsch, 2005).

With such widespread conflict around the world throughout history, it is likely that non-conflict disasters sometimes strike conflict zones. Examples of non-conflict disasters include earthquakes, industrial explosions, floods, transportation crashes, epidemics, and hurricanes. With conflict undermining governance, livelihoods, and basic services such as health, water, and sanitation, the overlapping conflict and non-conflict disasters are inextricably linked.

A war-weary population with reduced physical and psychological health is more susceptible to a pandemic. A government focusing on war might neglect promulgation, monitoring, and enforcement of earthquake-related building codes. Conflict frequently interferes with or cuts essential supplies such as food, medicine, and building materials, making it difficult for people to keep their homes and communities prepared for floods or storms.

These examples represent the well-known phenomenon of conflict increasing disaster vulnerability and exacerbating disaster impacts. Is the reverse possible? Could some form of non-conflict disaster striking a conflict zone lead to compassion, desire to help, or collaboration in order to deal with that disaster? This book covers some aspects of such 'disaster diplomacy'.

Basic definitions provide a starting point. As a starting definition, a disaster is 'a serious disruption of the functioning of a community or a society involving widespread human, material, economic or environmental losses and impacts, which exceeds the ability of the affected community or society to cope using its own resources' (UNISDR, 2009). Also a starting definition, diplomacy can refer to peaceful conduct of official business amongst sovereign-state governments and other government-related entities involved in world political matters (after Bull, 1977, although this definition is by no means exclusive to the English school of international relations).

From the definition of disaster, all forms of activities related to disasters can be described. Disaster-related activities refer to what happens – for example, investigations, proposals, policies, practices, and actions – before a disaster along with what happens after a disaster. Before a disaster, activities can cover prevention, planning, mitigation, preparedness, and risk reduction. Post-disaster activities incorporate response, relief, reconstruction, and recovery.

Many manuals attempt to put disaster-related activities into a sequence or a never-ending cycle. The cycle includes a disaster leading to post-disaster activities that then segue into pre-disaster activities until another disaster strikes. Most such models have been criticised as being culturally biased, unrealistic, or unhelpful for a long-term view. For example, Balamir (2005), Lewis (1999), van Niekerk (2007), and Wisner *et al.* (2004) indicate frustration with the cycle model and provide alternatives, plus La Red (La Red de Estudios Sociales en Prevención de Desastres en América Latina; www.desenredando.org) has long been debating how to do better than the 'cycle' concept.

The reason for the critiques is that, by having a disaster-orientated focus, the cycle could sometimes be interpreted as implying that disasters must always happen and that the disaster part of the disaster cycle can never be changed – whether or not that is the intention or implementation of the cycle. The critics of the 'disaster cycle' offer suggestions to explicitly move away from disaster, rather than always including it in the model. Their approach emphasises the importance of pre-disaster activities, rather than being reactive by putting measures in place to prevent the disaster that just happened.

No author denies the need to be ready for disasters and to implement appropriate post-disaster activities. They recognise that, while a world free from disasters would be ideal, current realities seem likely to preclude that for some time into the future. They also accept that it is equally unrealistic to assume that disaster must be an inevitable part of any efforts to deal with disasters.

Instead, disaster-related activities need to be seen as overlapping and connected phases, always with the goal of reducing disaster risk over the long term so that, even when a disaster occurs, it is not as bad as the previous time. That matches UNISDR's (2009) description of 'disaster risk reduction' as 'The concept and practice of reducing disaster risks through systematic efforts to analyse and manage the causal factors of disasters, including through reduced exposure to hazards, lessened vulnerability of people and property, wise management of land and the environment, and improved preparedness for adverse events'. Improvements are frequently seen in communities being able to help themselves (Ogawa *et al.*, 2005), in educational programmes (Holloway, 2009), and in legal frameworks for supporting disaster-risk reduction (Spence, 2004).

Understanding the 'disaster' part of disaster diplomacy must be seen within these wider contexts. All disaster-related activities are covered, pre-disaster activities and post-disaster activities, seeking to break the cycle.

The concept of 'diplomacy' also has wide contexts for disaster diplomacy, beyond the starting definition. Rather than referring strictly to bilateral or multilateral relations amongst entities representing sovereign-state governments and other government-related entities involved in world political matters, numerous aspects of international affairs, international relations, peace, and conflict are covered. For understanding 'diplomacy' within 'disaster diplomacy', those in conflict or collaborating (and sometimes both simultaneously) could be sovereign states, international organisations, non-profit groups, businesses, or non-sovereign territories. Many of the latter are sub-national jurisdictions with their own government such as municipalities or provinces. Scope further exists for individuals to be considered as the entities involved in disaster diplomacy irrespective of their affiliation with governmental or non-governmental bodies.

Bringing together 'disaster' and 'diplomacy' yields 'disaster diplomacy'. Is the topic worthy of study? The theme appears to be most popular in the media after a disaster hits a conflict zone or a country which has enemies. An expectation is often implied that disaster should bring peace, whether or not any precedent or realism exists for that expectation. Policy- and decision-makers can be forced to respond to populist pushing for a disaster diplomacy process that they would rather avoid – legitimately or otherwise.

Science tends to be more cautious about leaping into disaster-diplomacy desires or trying to force forward such a process. Instead, the question is asked: does evidence exist to support the hope for disaster diplomacy? By investigating multiple, diverse case studies, from numerous perspectives, this question is addressed at many levels in this book. In doing so, a balance is sought in the argumentation, seeing whether arguments favouring disaster diplomacy are justified – or whether so little empirical support for disaster diplomacy exists that it is a tenuous thread.

The disaster-diplomacy work reported in this book places heavy emphasis on relatively contemporary events and situations. The years 2000–10 are most prominent, although case studies from previous decades and centuries are mentioned. Most of the case studies, contemporary or older, are embedded in the long scientific and practitioner history that led up to the past decade or so of disaster diplomacy investigations.

The case studies and analyses, in fact, draw on the long history of many fields of study that involve disaster-related studies and diplomacy-related studies. All fields, all histories, all case studies, and all literature are not covered – far from it. To be comprehensive would require a volume several times the size of this one. Instead, this book provides a baseline for further exploration of the fields, the histories, the case studies, and the literature. With such a baseline – incorporating a clear inception point and a clear path for expanding knowledge based on previous work – disaster-diplomacy work for this book can be defined and summarised.

Disaster diplomacy examines how and why disaster-related activities do and do not reduce conflict and induce cooperation. Many variations of that definition, scope, and theme exist. This book explains how this summary evolved from principally one specific series of investigations (based on Kelman and Koukis, 2000), along with spin-offs, parallels, and tangents to that core statement.

That neither neglects nor precludes the impressive vastness of work relevant to disaster diplomacy that is separate from the work emphasised here. Rather, this approach achieves a balance of focusing the topic without constraining it, suitable for a single volume. It also assists in sifting evidence for and against disaster diplomacy, to ensure that recommendations can be supported by case studies. The manifestations of disaster diplomacy, the reasons for those manifestations, and the explanations regarding why more manifestations do not occur can now be examined.

1.2 A brief history of disaster diplomacy

Aspects of disaster diplomacy have long been studied and applied. Olson and Gawronski (2010) give a history of the literature on the politics of disaster, mainly from the perspective of the USA, that goes back to 1925. Similarly, Platt (1999) details the history of dealing with disasters in the context of democracy, again from the perspective of the USA.

For disasters across international borders, from the border towns of Eagle Pass, Texas, USA and Piedras Negras, Coahuila, Mexico, Clifford (1956) investigated the Rio Grande River's flood of 27–30 June 1954. His sociology study detailed patterns of individuals and organisations responding to the cross-border flood disaster. In addition to highlighting the importance of researching during a disaster situation, which was relatively unique at the time, Clifford's (1956) conclusions focus on the apparent porosity of the border with respect to social interactions. Both informal individual connections and formal institutional connections were seen as being important in helping both of the communities to respond to the disaster. In particular, 'The disaster acted as a catalyst, in a sense, which suddenly brought many areas of relationships into sharp focus' (Clifford, 1956: 137).

Twenty years later, Quarantelli and Dynes (1976) listed factors influencing, and consequences of, the presence and absence of different forms of post-disaster community conflict. Most case studies are American and the authors provide an American point of view regarding their expectations for cross-cultural differences and similarities. The study's conclusions are that most conflicts emerging in post-disaster activities were the same as the pre-disaster conflicts that existed within the community.

The (re)-emergence of these conflicts could be seen as evidence of the community recovering from the disaster. That can even 'reinforce the community cohesion which is produced during the emergency period' (Quarantelli and Dynes, 1976: 149). This post-disaster community cohesion along with improved morale due to the disaster and the disaster response are identified in the case studies. Yet the disasters examined do not appear to have had excessive mortality, widespread devastation across the entire community, or more than one community group in intense or violent conflict with many other community groups.

At the same time and from an international perspective, Glantz (1976) delved into disaster politics. Based on drought in the Sahel, his book explains how politics influence why disasters happen and how disasters could be averted. The focus is on disaster-related myths and tackling those myths. Relevant to disaster diplomacy, one myth suggested is that 'Things will change' after a disaster. The reality is that improvements and aims for positive change are often hindered by governance inertia. Glantz (1976) sets a baseline for examining misconceptions about the interaction of politics and disaster and how to overcome those fallacies.

Fast forwarding more than 20 years, Lewis (1999) covers several disaster-diplomacy examples. He notes how 'The November 1970 cyclone, and its subsequent alleged mismanagement, was one of the many influences that triggered the Bangladesh War of Independence which commenced in March 1971' (Lewis, 1999: 25). He describes how the 1972 Managua earthquake helped the Sandinista rebellion and civil war to gain traction that, seven years later, overthrew the dictatorship running Nicaragua. De Boer and Sanders (2005) provide details on that case study, emphasising the obvious post-disaster governmental corruption that drew powerful support to the previously marginalised Sandinista movement.

Nel and Righarts (2008) summarise and provide earlier references for a long history of disasters impacting conflict. They start with the earthquake that devastated Sparta in 465/464 BC providing the spark for a slave uprising. De Boer and Sanders (2004) cover several volcanic case studies changing physical and political landscapes, from the eruption of Thera, Greece in the Bronze Age to the 1902 eruption of Mount Pelée in Martinique in the Caribbean. Broder *et al.* (2002) discuss diseases decimating armies at war from Alexander the Great to World War II.

Olson and Drury (1997) and Drury and Olson (1998) are examples of quantitative investigations from the long history of disaster-conflict studies. They statistically analysed data between 1966 and 1980 to seek connections between disasters and political unrest. To be included, disasters had to kill more than 1,500 people during the study period. Other data for each case study needed to be available for two years prior to the disaster and for seven years after it. That produced a sample of Bangladesh, China, Guatemala, Honduras, India, Iran, Nicaragua, Nigeria, Pakistan, Peru, the Philippines, and Turkey.

Olson and Drury (1997) showed that worse disasters can lead to more political unrest, especially by creating an opportunity for critics to tackle a regime. In that sense, disasters 'can be seen as democratizing in some respects' (Olson and Drury, 1997). In parallel, using a combination of political and developmental variables, Drury and Olson (1998) concluded that increasing development, increasing income equality, and increasing regime repressiveness all led to less post-disaster political strife.

Even though the interactions amongst disasters and the topics of diplomacy, peace, and conflict were being researched and published on, the phrase 'disaster diplomacy' seems to appear infrequently prior to 2000. In a practitioner piece on communications for disaster response, Silverstein (1991) states only that 'Communications networks are the routes of international disaster diplomacy' (p. 3) without further discussion or analysis.

In the purely academic literature, Dove and Khan (1995) quote the phrase 'disaster diplomacy' from a media article of that title, a 1991 piece in the 'Dhaka Courier' from Bangladesh, but they do not describe or analyse the phrase's meaning. Dove (1998) quotes the same article, briefly discussing it in the context of how the Bangladeshi media and government frame disasters in their country.

Then, Kelman and Koukis (2000) started an effort to systematically document and analyse disaster diplomacy through editing four papers. The papers are Ker-Lindsay (2000) examining the Greece–Turkey case study, Glantz (2000) detailing Cuba–USA, Holloway (2000) covering Southern Africa during the 1991–93 drought, and Comfort (2000) completing a comparative analysis of the three case studies using a complex adaptive systems framework. Each of these papers is detailed in later sections of this book.

The publication by Kelman and Koukis (2000) was originated by Theo Koukis who proposed a special section of the journal *Cambridge Review of International Affairs*. His suggestion was based on the ongoing Greek–Turkish rapprochement that appeared to result from the earthquake in Turkey in August 1999 and the earthquake in Greece in September 1999. From media discussions, Koukis suggested the title of 'Earthquake Diplomacy'. The journal's editor at the time, Charlotte Lindberg Clausen (now Charlotte Warakaulle), suggested expanding the section's scope and title to 'Disaster Diplomacy'. I then joined the team editing the 'Disaster Diplomacy' section, leading to the publication of Kelman and Koukis (2000). Based on this work and the interest generated by the publication, in January 2001 I launched the disaster-diplomacy website now at www.disasterdiplomacy.org.

2 Moving forward with disaster diplomacy

2.1 What this volume offers

From the origins described in Chapter 1, many more disaster-diplomacy case studies have been examined. These investigations involved numerous authors, interests, and disciplinary approaches. This book details and analyses that work. The focus, although not exclusively, is on one particular dimension of 'disaster diplomacy': that of the work over the past ten years or so that is related to Kelman and Koukis (2000).

This focus in no way denigrates, dismisses, or ignores pre-2000 work or post-2000 publications covering overlapping topics without disaster diplomacy as the core. In contrast, this book needs, uses, interprets, and references some (actually a small part) of that work. But that material is not the principal part of this volume, nor is that material addressed comprehensively. Any publication must scope itself. This book does so by limiting itself to disaster-diplomacy work from mainly 2000 to 2010. This scoping also assists in ensuring this book's originality, rather than repeating the deep, extensive, and expansive work, as referenced throughout this book, published on similar topics, but outside the 2000–2010 disaster-diplomacy focus.

The interpretation and analysis here is also tilted towards the hope that research findings, fully evidenced and analysed, could assist policy makers and decision makers – at all levels, from international governance to people going about their day-to-day lives. As such, many sides of arguments are presented and different formulations of questions are provided, because all of those appear in the literature, forming the history of the topic and providing the basis for the current understanding. Selecting evidence to promote only one argument or avoiding references that contradict each other cannot give the full picture. The aim is that, as more science factoring in all sides of a debate is applied to policies and decisions, those policies and decisions will improve by having considered all viewpoints rather than focusing on just one – especially one matching pre-conceived notions.

Is that approach naïve? Policy- and decision-makers tend to want summaries and can be adept at referencing literature that supports their already established viewpoints. The ethos of trying to ensure that balanced, accurate, and precise science assists policies and decisions has numerous names and interpretations in the literature, with two examples being 'useable science' (Glantz, 1997) and 'people's science' (Wisner et al., 1977). Each approach has met with successes and failures. This book, too, could be accepted by those with time to sift through the various arguments while being rejected by those who want one opinion only or who have already made up their mind, irrespective of the evidence – or lack thereof.

While applied science that can be used by non-scientists emerges from this book, fundamentally this volume aims for the highest academic standards. As such, the purpose of this book is to provide a scientific summary and analysis of recent disaster-diplomacy work to systematically understand and explain disaster diplomacy. Rather than accepting media or practitioner constructs of disaster diplomacy, or assuming what must occur with respect to disaster diplomacy, this book seeks and weighs evidence, trying to be certain of what that evidence states before reaching conclusions.

The book is aimed primarily at applied researchers along with policy- and decision-makers who seek to use science related to disasters and politics. The secondary readership is policy- and decision-makers focusing on international NGOs, the UN, and aid agencies, as well as graduate students taking courses or completing dissertations in this topic.

2.2 What this volume does not offer

Being an academic book, plenty of reference is given to past work while the discussion builds on that past work. Only certain areas of past work are covered. The discussion here is not steeped in heavily theoretical approaches, especially different forms of international relations or international affairs theories.

As one prominent example, the discussions amongst the political realism school, the English school, and others are not factored into the analyses in this book. Various forms of environmental conflict theory are touched upon only superficially, without detailing the extensive literature of principal authors (for example, Homer-Dixon, 1999), those who critique such work (for example, Hartmann, 2010), and the debates amongst them ('Exchange', 2003; Kahl, 2002). Democratic peace theory (for example, Doyle, 1983a, 1983b) and its critiques (for example, Rosato, 2003) are not addressed along with parallel theories and applications. Many other approaches and debates in the literature covering peace, conflict, and diplomacy are absent from this volume.

This book's bias is very much on the disaster side of 'disaster diplomacy'. The applied-science literature from disasters, development, and sustainability is covered much more than that from political science, international relations, international affairs, conflict resolution, and peace and conflict studies. Rather than disciplinary approaches thick in academic discourse, the interdisciplinary – often non-disciplinary – field of disaster research is highlighted at the expense of other fields. Much of this bias emerges from this book's principal scope, narrowed to examining approximately a decade of work that has a basis in one publication, Kelman and Koukis (2000).

With that basis, the analysis here is almost exclusively qualitative. Some quantitative studies are addressed in section 5.1, yet for the most part, quantitative analytical methods are not used here. That is partly due to the limitations of quantitative data, as discussed in section 5.1. Nevertheless, many quantitative studies have done well in admitting honestly, and overcoming, many of the main data limitations (for example, Gleditsch *et al.*, 2002; Harbom and Wallensteen, 2009; Lacina and Gleditsch, 2005). These and other studies are drawn upon in appropriate places. Nonetheless, the decade of disaster diplomacy based on Kelman and Koukis (2000) has contributed mainly qualitative studies to the literature, so that aspect is emphasised in this book.

Hopefully, the reader will be inspired to make connections beyond what this book offers, ensuring that future work builds on the material here to fill in the gaps – and especially to do better than what could be offered here.

2.3 The structure of this volume

Disaster diplomacy started mainly with hypotheses explored through case studies that draw on empirical evidence from the authors' own experiences yielding Glantz (2000), Holloway (2000), and Ker-Lindsay (2000). Then, theoretical developments and explanations were analysed from the case study material, as demonstrated by Comfort (2000). The feedback between field evidence and theoretical developments continues.

This book's structure emerges from that evolution. Following the background and scoping provided in Chapters 1 and 2, Chapter 3 presents disaster diplomacy hypotheses and research questions. Multiple forms are provided in order to cover what has appeared in the literature, illustrating the evolution of the concepts to the current view. Chapter 4 then presents a variety of case studies with limited connection amongst them to test the hypotheses and to consider the research questions from many angles, aiming to ensure that various perspectives are given due consideration.

With that basis for investigation, the hypotheses and research questions (Chapter 3) are examined in light of the case studies (Chapter 4) through Chapters 5, 6, and 7 that seek to explain disaster diplomacy. The explanations revolve around why and how disaster diplomacy does and does not manifest, often covering the case studies from different perspectives to be certain of having tested the available evidence before drawing conclusions. Chapter 5 covers and critiques quantitative approaches followed by qualitative typologies for the case studies. Chapters 6 and 7 take those qualitative typologies and formulate a suite of pathways describing how and why disaster diplomacy succeeds, is made to succeed, fails, and is made to fail.

10 *Moving forward with disaster diplomacy*

With that explanatory and predictive foundation – or lack thereof – for disaster diplomacy, further case studies are examined through disaster diplomacy spin-offs in Chapter 8. Chapter 9 summarises limitations of disaster-diplomacy research and application. These are ethics considerations, confounding factors affecting the analyses, and potential biases in those who explore or try to use disaster diplomacy.

Then, disaster-diplomacy lessons are extracted for Chapter 10. Chapter 11 details future work, to indicate the gaps that remain in understanding and applying disaster diplomacy and how those gaps could be overcome. Chapter 12 sums up, balancing the wider perspectives given in Chapter 1's title 'The Origins of Disaster Diplomacy' by indicating 'The Future of Disaster Diplomacy'.

3 Hypotheses and research questions

3.1 Definitions of disaster diplomacy

Kelman and Koukis (2000) started with the question 'Do natural disasters induce international cooperation amongst countries that have traditionally been "enemies"'? Two limitations of that initial question are immediately evident: the terms 'natural disasters' and 'international cooperation'.

The term 'natural disaster' turned out to be a poor choice, due to the connotation that the disaster is caused by nature or that these disasters are naturally what happens when society interacts with the environment. In contrast, disaster-related research and practice has long been finding societal rather than natural or environmental causes for disasters.

One of the earlier discussions of this viewpoint followed the 1 November 1755 earthquake and tsunami that hit Lisbon, Portugal. Rousseau (1756) wrote to Voltaire describing his ideas that the houses which collapsed were not built by nature and that an earthquake that did not affect human settlements would not be a disaster. He also noted that Lisbon's high population density seemed to be a factor in the high death toll. Rousseau further explained how he believed that many unnecessary injuries and deaths resulted due to people's poor reactions following the earthquake.

Over two centuries later, O'Keefe et al. (1976) published a paper titled 'Taking the "naturalness" out of natural disasters', describing how human behaviour leads to 'natural disasters'. They explained that disasters were increasing, but not due to nature changing. Instead, 'the growing vulnerability of the population to extreme physical events' (O'Keefe et al., 1976: 566) was leading to the observed increase in disasters. The next year, Tiranti (1977) discussed 'The Un-natural Disasters'.

The argument is that few natural disasters exist because most disasters require human input. Nature sometimes provides input through a normal and necessary environmental event, such as a flood or volcanic eruption. But human decisions have put people and property in harm's way without adequate measures to deal with the environment in which they live. The conclusion is that those human decisions are the root causes of disasters, not the environmental phenomena. Disasters are not 'natural', but are social constructions.

Disaster science (for example, Hewitt, 1997; Lewis, 1999; Mileti et al., 1999; Oliver-Smith, 1986; Steinberg, 2000; Wisner et al., 2004) now fully accepts that human actions, behaviour, decisions, and values create vulnerabilities and those vulnerabilities are the fundamental cause of disasters. Hence, disasters tend not to be 'natural', either literally or figuratively.

Even meteorite strikes are not natural because society has established methods for monitoring space, identifying potential threats, and countering them (for example, Carusi et al., 2005; Cellino et al., 2006; Peter et al., 2004; Price and Egan, 2001; Stokes et al., 2000). Society makes a choice not to spend the money to do so, despite knowing the potential implications of not providing enough resources to deal properly with the identified problem. Some exceptions exist, such as basaltic flood eruptions and nearby supernova explosions which have no known mechanisms for preventing the hazard or for reducing global-scale vulnerability.

While such debates might seem to be overly academic, practitioners accept the notion that human input exists to all disasters. Abramovitz's (2001) report 'Unnatural Disasters' describes the factors which make disasters with environmental phenomena unnatural: 'undermining the health and resilience of nature, putting ourselves in harm's way, and delaying mitigation measures'. Turcios (2001) asserts 'Natural disasters do not exist; they are socially constructed'. UNISDR (2002) notes that 'Strictly speaking, there are no such things as natural disasters' while UNISDR's (2004, 2009) terminology of basic disaster-risk reduction terms does not include 'natural disaster'.

Even the term 'disaster' in 'disaster diplomacy' is restrictive, because it implies that a disaster is necessary for disaster diplomacy to be considered. As noted in the case studies (Chapter 4), much potential cooperation and peace building is seen around pre-disaster activities. Examples include Cuban and American scientists cooperating to monitor hurricanes prior to landfall (section 4.6), cooperation across southern Africa in 1991–93 to avert a drought emergency becoming a drought disaster (section 4.4), and international collaboration for a tsunami detection and warning system. Another example is the Middle East Regional Cooperation Program funded by the United States Agency for International Development which brings together Jordan, Israel, and the Palestinian National Authority to work on earthquake hazard assessments for building codes to incorporate earthquake resistance.

To avoid restrictions from the term 'disaster', disaster diplomacy was defined to expand beyond the occurrence of a disaster to encompass all forms of disaster-related activities. While disasters in conflict zones continued to garner the most attention and the most case studies, the explorations included other collaborative efforts for disaster-risk reduction activities such as warning systems, building codes (including monitoring and enforcement), and land-use planning for disaster-risk reduction along borders.

The second main concern with the initial question posed by Kelman and Koukis (2000) is the term 'international cooperation'. Assuming that only international diplomacy is relevant is narrow. Examples that have been explored beyond international diplomacy are intra-state conflict resolution in Aceh (section 4.10) and the Philippines (section 4.3) along with disaster diplomacy involving sub-national jurisdictions dealing with their governing state or other sovereign states (section 8.2).

Based on these expansions, the current statement defining disaster-diplomacy work is now that disaster diplomacy investigates how and why disaster-related activities do and do not induce cooperation amongst enemies. Previous variations include 'Could disaster-related activities, pre-disaster and post-disaster, positively affect relations amongst states which would not normally be prone to such cooperation?' (Kelman, 2006b, p. 70) and 'How and why disaster-related activities do and do not yield diplomatic gains, looking mainly at disaster-related activities affecting diplomacy rather than the reverse' (Gaillard *et al.*, 2008, pp. 511–12).

Difficulties with the term 'enemies' have also been articulated. That term can be seen to be too strong, in terms of suggesting violent conflict or complete antipathy towards each other. Yet friends and allies can have disagreements and conflicts, yielding disaster-diplomacy possibilities without the parties necessarily being full or violent enemies. Canada and the USA are not usually considered to be enemies, yet the Red River Floods in 1997 are an example of the conflict-related successes and failures that can occur for countries dealing with cross-border disasters (Wachtendorf, 2000).

More speculatively, Kelman (2007) considered the possibility that Hurricane Katrina (section 4.12) could assist in solving the long-running and often bitter lumber-trade dispute between Canada and the USA. With much of the forests along the USA's Gulf Coast devastated by the disaster, would the USA try to reach a lumber deal with Canada to assist post-hurricane reconstruction? As with most disaster-diplomacy case studies, nothing happened.

In making the definitions of 'disaster' and 'diplomacy' more encompassing, a further shift in the defining statement appeared as more disaster-diplomacy literature was published. Kelman and Koukis (2000) straightforwardly asked whether or not disaster diplomacy is witnessed. That could be answered with 'yes' or 'no'. As disaster-diplomacy work evolved, the question became 'how and why'. That explicitly explores the ongoing processes that are observed and seeks explanations for those observations.

Additionally, 'do and do not' was included to ensure that no assumption was made that disaster diplomacy would (or would not) be successful. Overall, the disaster-diplomacy question has broadened, to look at successes and failures, and has deepened, seeking to explain 'how' and 'why' rather than just identifying 'yes' or 'no'. Disaster diplomacy investigates how and why disaster-related activities do and do not induce cooperation amongst enemies.

3.2 Hypothesis: catalysis, not creation

At the beginning of disaster-diplomacy investigations through Kelman and Koukis (2000), the disaster-diplomacy hypothesis was frequently that disaster-related activities should or will support diplomatic efforts. That hypothesis was rarely confirmed by the investigations. Invoking the opposite hypothesis, that disaster-related activities should not or will not support diplomatic efforts, could also not be proven. The reason is that clear elements of disaster diplomacy are frequently witnessed.

The hypothesis was modified to suggest that disaster diplomacy has a tangible, but not an overriding, presence. Disaster-related activities can influence, support, affect, push along, or inhibit diplomatic processes, but that does not always occur. Furthermore, disaster-related activities support diplomacy only when the diplomacy has already started with a basis other than the disaster-related activities. Disaster-related activities alone have not been shown to produce new diplomatic results or initiatives. Where disaster-related activities are affecting diplomatic processes, any gains can be easily and rapidly lost due to other events or due to factors influencing the diplomacy. Those other events and factors are sometimes further disasters or disaster-related activities.

That discussion provides the main disaster-diplomacy hypothesis that disaster-related activities can act as a catalyst, but not as a creator, of diplomacy, although catalysis is not always seen. Disaster-related activities do not create fresh diplomatic opportunities, but they have the possibility of catalysing diplomatic action. That possibility is not always fulfilled. Three sub-points emerge from this disaster diplomacy hypothesis.

First, in the short-term – considering a time-scale on the order of weeks and months – disaster-related activities can, but do not always, impact diplomacy. Where disaster-related activities do influence diplomatic activities, especially by spurring them on and supporting them, a pre-existing basis is needed on which to found that diplomacy. That pre-existing basis could be trade links, cultural connections, secret or open peace negotiations amongst the parties in conflict, or joint sports events.

Second, over the long-term – considering a time-scale over years – non-disaster factors are generally seen to have a more significant impact on diplomatic processes than disaster-related activities. Examples of non-disaster factors are distrust, changes in leadership which could be political or civil depending on who the real power brokers are, or belief that a past conflict or grievance (even if centuries ago) should supersede present-day disaster-related and peace-related interests. Overall, these factors refer to political priorities for action that are different from reducing conflict and gaining diplomatic dividends. That can be as simple as holding onto power irrespective of the costs, whether that be through re-election or less democratic means.

Third, evidence exists for disaster-related activities sometimes having the opposite outcome to disaster diplomacy. That is, disaster-related activities can sometimes exacerbate conflict, reduce diplomacy, and diminish peace prospects. Examples are fighting over disaster relief supplies, using disaster-related activities as a ploy to gain or retain power, or using a disaster to undermine those in power by accusing them of being at fault for the disaster. Another possibility is using an enemy's moment of weakness or distraction in dealing with a disaster to gain political, diplomatic, or military advantages.

In summary, the disaster-diplomacy hypothesis and three sub-points lead to the political truism that disaster-related activities are not usually a high priority for decision-makers. Disasters sometimes garner attention and resources, increasing the prominence of disaster risk reduction before a disaster manifests and leading to legitimate action. Why that occasionally but not usually occurs is an ongoing challenge to explain (for example, Lewis, 2003; Wisner, 2003). For disaster diplomacy, investigations need to further break down the elements within the hypothesis to start developing a framework of analysis or a checklist to be used for comparing case studies and for explaining the reasons underlying the hypothesis.

3.3 Questions for disaster diplomacy

To develop a framework of analysis for examining the disaster-diplomacy hypothesis (section 3.2), Kelman (2006a) provided a checklist of four yes/no questions. The questions were to be asked for each case study to determine whether or not it represents disaster diplomacy and, if so, to what degree. The questions are:

1 Did disaster-related activities lead to diplomatic interactions?
2 Did new diplomacy emerge? If disaster-related activities influence ongoing diplomacy, then catalysis, not creation, of diplomacy occurs.
3 Is the diplomacy legitimate? The parties involved must be seeking rapprochement rather than using disaster-related activities for public relations – or waiting for an opportunity to bail out of the diplomatic process. Assessing legitimacy is highly subjective, especially since the parties involved might change their legitimacy interests over time.
4 Does the diplomacy last? The time-scale for which diplomacy could be said to be 'lasting' is also subjective. Kelman (2006a) suggested that years would be an appropriate time-scale. Months or weeks was deemed to be too short to ensure that long-term diplomatic results are achieved. Additionally, given that disaster-diplomacy case studies so far show that disaster-related activities sometimes influence diplomacy over weeks and months, but then it tends to fade away, a longer time-frame would make sense to seek truly long-term disaster diplomacy. Considering a time-frame beyond years, a time-scale of decades is difficult to analyse. Such investigations would naturally be precluded for recent case studies. Cuba–USA (section 4.6) is an example of a case study that has effectively been analysed over decades (Glantz, 2000) because some of the background is relevant back to Fidel Castro's 1959 seizure of power and falling out with the USA over the next few years. For material on Castro's governance of Cuba, see for example Gonzalez and Ronfeldt (1986) and Miller and Kenedi (2003).

16 Hypotheses and research questions

In considering these questions for post-tsunami Aceh (section 4.10), Gaillard *et al.* (2008) expanded the checklist to include:

5 Is the diplomacy dependent on the characteristics of the post-disaster reconstruction? Could a reconstruction process that does not address longer-term political and economic concerns hamper or undermine a peace process?
6 How does post-disaster diplomacy address those long-standing political and economic issues?

By investigating the long-term development and vulnerability issues that contributed to creating both the conflict and the tsunami vulnerability in Aceh, these two additional questions ensure that a time-frame of decades is adopted for analysing the post-disaster diplomacy.

Gaillard *et al.* (2009) recognised that the set of questions answered for post-tsunami Aceh in Gaillard *et al.* (2008) were formulated around knowledge that was specific to that case study. To make the questions more generic, and to apply to the case study of the Philippines and the USA after the 1991 eruption of Mt Pinatubo, Gaillard *et al.* (2009) updated the question checklist to:

1 Did disaster-related activities influence diplomatic activities? If not, then by definition it is not a disaster-diplomacy case study.
2 Did new diplomacy result? If disaster-related activities influence an already-established diplomatic process, then catalysis rather than creation occurred.
3 Is the diplomacy legitimate? Legitimacy refers to the parties involved in the conflict seeking rapprochement rather than using the events as a public relations exercise or awaiting an opportunity to avoid proceeding further with the diplomacy.
4 How long does the connection between the disaster-related and diplomatic activities last?

If a disaster has occurred:

5 How closely is post-disaster reconstruction linked to diplomacy?
6 How well does post-disaster diplomacy address long-standing political topics and livelihood concerns? Examples of political topics are empowerment, participation in governance and the right to vote freely and fairly. Examples of livelihood concerns are getting enough to eat, having diverse livelihood choices, and generating possibilities for developing further livelihood options such as through education and training.

The questions are consolidated and updated here, to make them more generic and open-ended. That way, they could apply to any form of case study while encouraging deeper investigation than yes/no answers. Based on the history and discussion given above, the disaster-diplomacy research questions are now:

1 How are disaster-related activities influencing diplomatic activities?
2 To what degree are those diplomatic activities new and to what degree were they ongoing?
3 How legitimate do the parties involved view the diplomacy? How are the parties trying to make the diplomacy succeed or fail?
4 How long does the connection between the disaster-related and diplomatic activities last? Why does the connection persist or fade away?
5 How well does the disaster diplomacy address long-standing political and livelihood concerns, particularly related to vulnerability?

4 Empirical evidence
Case studies

4.1 Organising case studies

Disaster-diplomacy case studies appear in various levels of detail. They range from in-depth, published research involving extensive field work through to ideas or informal anecdotes with limited evidence, often just asking questions to be investigated or awaiting a formal literature search. A balance also exists with case studies constructed from media accounts and those that have been subject to proper academic analysis. That means that the case studies are of many forms (Table 4.1).

Some cover a specific geographical area over a specific time period, involving many disaster-related activities. Others are focused on a specific disaster, such as a named tropical cyclone or preparedness for a specific earthquake fault failure and failure type. Some case studies are a particular disaster type that affects numerous locations around the world, sometimes being tackled by international institutions. Other case studies cover a specific disaster topic, not linked to a specific disaster type.

This chapter describes many of the case studies that have been investigated or identified with regards to disaster diplomacy (Table 4.1). The material provides the empirical basis for the later, more theoretical discussions and analyses of how and why disaster diplomacy appears and does not appear in different forms. As such, limited analysis is presented in many of the case studies in this chapter. Instead, the text of this chapter tends to be more focused on case-study histories, sequences of events, and referenced literature.

Organising the case studies into groups or clusters is not straightforward. Organising by geographic region could exclude case studies spanning continents such as Iran–USA and vaccination programmes. Organising by disaster-diplomacy outcome could only be achieved after the evidence is presented and analysed for each case study. That is completed in subsequent chapters. Organising by hazard or specific disaster type might not fully acknowledge the role that vulnerabilities and capacities play in disaster-related activities. Additionally, several case studies – such as North Korea, the Philippines, and disaster-casualty identification – involve several, disparate hazard types. In the end, the case studies need to be presented in a linear

Table 4.1 Disaster diplomacy case studies in this chapter (in alphabetical order by case-study name)

	Case-study type	Main sources
Burma and Cyclone Nargis	Specific disaster	Media
Casualty identification	Disaster topic	Academic references
China and the 2008 earthquake	Specific disaster	Media
Cuba–USA	Geographical area	Academic references
Ethiopia–Eritrea	Specific disaster	Media
Greece–Turkey	Geographical area although focused mainly on two specific disasters	Academic references
Hurricane Katrina	Specific disaster	Academic references
India–Pakistan, 2001	Specific disaster	Academic references
India–Pakistan, 2005	Specific disaster	Media
Iran–USA	Geographical area, although focused on mainly one specific disaster	Academic references
North Korea	Geographical area	Media
Philippines	Geographical area	Media
Sea-level rise and island evacuation	Disaster type	Academic references
Southern Africa 1991–93	Specific disaster	Academic references
Tsunami, Aceh	Specific disaster, although focused on mainly one specific geographic area	Academic references
Tsunami, Sri Lanka	Specific disaster, although focused on mainly one specific geographic area	Academic references
Tsunami, other case studies	Specific disaster, although focused by specific geographic area	Media
Vaccination programmes	Disaster type	Academic references

sequence, with the decision being more or less arbitrary, such as Table 4.1 alphabeticising by case-study name.

To bring a modicum of coherence to the order of case studies, they are presented in this chapter with specific geographical locations appearing first, in chronological order from the case study's approximate start time, without forgetting that most case studies have a long history that goes back decades, sometimes centuries. Then, the final three case studies cover long-term, ongoing disaster-related topics: the creeping environmental change of sea-level rise due to contemporary climate change, disaster-casualty identification, and international vaccination programmes. The final section foreshadows the collective disaster-diplomacy lessons from the presented case studies.

This structure produces a long chapter with extensive cross-referencing amongst the case studies. The diversity of material is also evident, embracing

20 *Empirical evidence*

all forms of disaster-related activities around the world with diplomatic interactions entailing numerous forms of activities. This level of detail and variety of disaster and diplomacy activities provides a solid empirical basis for exploring disaster diplomacy, rather than selecting case studies that suit a pre-formed conclusion.

4.2 Iran–USA from 1990 onwards

Iran and the USA have had tense relations for decades. From World War II until 1979, the USA supported the Shah of Iran in governing the country through an oppressive dictatorship that consolidated its power by overthrowing a democratically elected government in 1953. The two main reasons for American support of the regime were to counter communist influence and to gain access to oil. A series of grassroots demonstrations and massacres of the demonstrators in 1978 led to the Shah fleeing Iran in 1979. Soon after, Iran heralded the return of Ayatollah Ruhollah Khomeini, a religious leader, who installed his own oppressive, but explicitly anti-American, dictatorship.

In November 1979, Iranian students took over the American embassy in Tehran, holding several dozen diplomats hostage. They were eventually released in January 1981. Meanwhile, a war had started between Iran and Iraq with the USA supporting the brutal dictatorship of Saddam Hussein in Baghdad, in order to try to weaken Iran's influence. The war lasted until 1988 and killed over 500,000 people, including the fatalities from Iraq's use of chemical weapons. The Iraq–USA alliance also permitted American access to Iraqi oil supplies while leading to an increasing American military presence in and around the Persian Gulf and Gulf of Oman. In July 1988, the American warship the USS Vincennes shot down commercial airline flight Iran Air Flight 655 killing all 290 passengers and crew on board.

Throughout this time period, religious fundamentalism in the region continued to take a decidedly anti-American and anti-Western tone, which resonated from and with Iran's approach to governing. American support for monarchical dictatorships in Saudi Arabia and Kuwait, mainly for attempting to limit the influence of religious fundamentalism and for protecting oil supplies, did not endear many of the region's leaders to American stances. None of that stopped secret deals between Tehran and Washington, DC in the 1980s regarding the USA selling weapons to Iran in exchange for freeing American hostages in Lebanon.

In the context of the decline of power of the USSR and the communist Eastern Bloc in the late 1980s, Iraq under Saddam Hussein's leadership invaded Kuwait in August 1990. His military forces were expelled by American-led forces in February 1991, but Saddam Hussein was permitted to retain power in Baghdad. The post-war years brought accusations of Saddam Hussein's attempts to retain and seek more biological, chemical, and nuclear weapons leading to bitter confrontations with UN inspectors in the country. That was eventually one of the excuses used by the USA and the UK to attack Iraq in 2003, deposing Saddam Hussein.

In parallel, during these years, Iran also came under UN criticism for alleged attempts at producing nuclear weapons. As with Iraq, diplomatic fights with international weapons inspectors and with the UN continued over several years. There was open speculation that either the USA or Israel might launch a military strike against Iranian nuclear-energy facilities. Throughout this period, numerous earthquake disasters struck Iran.

On 21 June 1990, an earthquake killed over 35,000 people in Gilan, Iran, on the Caspian Sea. Iran's government declined many offers of external assistance, including from the USA, and was especially trying to avoid help from the West (Warnaar, 2005). Nevertheless, within days, the American Red Cross and AmeriCare were providing help in Iran. With support from the USA's government, a private American relief plane landed with supplies for the non-governmental organisations. No political links were attempted or were made and no impact on Iran–USA relations was evident.

Almost 12 years to the day, on 22 June 2002, another earthquake hit nearby, killing several hundred people. As discussed with respect to Hurricane Katrina (section 4.12), months earlier, the US President had labelled Iran part of the 'Axis of Evil' (Bush, 2002) for supporting terrorism and for opposing American interests. Nevertheless, through complex negotiations, aid from the USA's government arrived in Iran indirectly on 2 July 2002. President Bush stated that 'Human suffering knows no political boundaries' eliciting a response from Tehran that the aid had 'no political character'. No discernible impact on Iran–USA relations was noted. Both countries seemed to want to cooperate with regards to the disaster as long as no diplomatic outcomes would result.

Then, on 26 December 2003, an earthquake struck southern Iran with most destruction witnessed in the UNESCO World Heritage city of Bam. More than 25,000 people were killed with initial estimates of the death toll being above 40,000. In the following days, the USA offered aid which was accepted by Iran, including the use of American aircraft to deliver supplies and personnel. A media frenzy related to disaster diplomacy and earthquake diplomacy resulted.

As Cater (2003) pointed out, the potential disaster diplomacy in this case was not just with regards to the USA. Egypt did not have diplomatic ties to Iran, but President Hosni Mubarak extended condolences and delivered aid.

On the same day that Cater (2003) published his insightful analysis, 30 December, US Secretary of State Colin Powell remarked that 'We should keep open the possibility of dialogue at an appropriate point in the future' with regards to the American assistance to Iran. That led some to believe that his government's hard line against Iran was softening because of the earthquake. But in early January 2004, Iran soon rejected the Bush administration's offer to send a high-level delegation, which would have included Senator Elizabeth Dole, a Republican from North Carolina.

It appears as if the White House made a faux pas by announcing the American delegation before clearing it with Iran through diplomatic channels. In commenting on the disaster-diplomacy speculation, Iran's President Mohammad Khatami said 'Humanitarian issues should not be intertwined with deep and chronic political problems. If we see a change both in the tone and behaviour of the U.S. administration, then a new situation will develop in our relations.' That reaction demonstrates the desire to separate the politics and the disaster cooperation.

Furthermore, Powell's statement needs to be viewed within contexts wider than the earthquake and wider than that final week of 2003 where Iran and the USA seemed to be coming closer together. Powell's view towards Iran at the time placed potential rapprochement in the context of Iran's agreement to permit UN inspections of Iran's nuclear-energy facilities in addition to moves towards reconciliation with Egypt and Jordan. Jordan, like Egypt, had offered earthquake-related assistance to Iran. The concern was not just Iran–USA after the earthquake, but an international re-engagement with Iran on multiple issues.

In fact, in October 2003, two months before the Bam earthquake, US Deputy Secretary of State Richard Armitage told the Senate Foreign Relations Committee that 'We are prepared to engage in limited discussions with the government of Iran ... as appropriate' (Armitage, 2003). He further emphasised statements from Bush and Powell earlier in October that downplayed the possibility of using force against Iran, in direct contrast to what had happened in Iraq.

The USA's earthquake relief to Iran was likely acceptable to both sides due to this ongoing process of diplomacy, rather than the earthquake relief opening up new diplomacy. The USA's government's post-earthquake stance to Iran was perfectly consistent with its pre-earthquake stance. Yet after the media seized on Powell's 30 December statement as an example of disaster diplomacy, despite the statement's vagueness, both Bush and Powell were forced to clarify that their policy had not changed. Some saw that as backtracking. Conversely, 'no change' was a legitimate attitude given that Powell's 30 December statement said nothing different from the USA's government's October statements.

Meanwhile, as shown by the statement from Iran's President and Iran's rejection of a high-level American delegation, Iran seemed to aim to avoid any earthquake-related rapprochement. Javad Zarif, Iran's UN ambassador, echoed these sentiments: 'We appreciate the importance of the humanitarian gesture ... the United States said this is for humanitarian purposes, and that is how we have taken it.' That echoes the quotation given above from Iran's President. The lack of disaster diplomacy in this case study is further supported by Iran's immediate post-earthquake declaration that Israeli aid would not be accepted. That is despite Israel's long history of sending search-and-rescue teams to disasters in other countries.

As January 2004 progressed and the earthquake and possible earthquake diplomacy faded from the media and diplomatic spotlights, politicians in both countries started looking towards their future. Parliamentary elections were due in Iran in February while November would bring Presidential and Congressional elections to the USA. The incumbents in both countries were seen as hardliners in dealing with 'enemies'. To avoid weakening their appeal to their power base, any sign of conciliation could be turned against them in the elections. That was to be avoided by both sides, even if that meant sacrificing possible disaster diplomacy from the Bam earthquake.

Why, then, would the USA offer aid to Iran and why would Iran accept that aid? Several linked reasons can be explored, as summarised from Warnaar (2005) and Kelman (2007), indicating the complexities involved in disasters, diplomacy, and disaster diplomacy.

First, Iran recognised and admitted that extensive foreign aid would be needed to deal with the catastrophe. One result was that aid from Egypt, Jordan, and the USA became acceptable, even though any form of aid from Israel was still deemed to be unacceptable. Iran's leadership seemed to have had personal antipathy towards Israel, plus Israel was presumably seen as more of an enemy than the other countries. An inadequate response to the disaster due to lack of aid, or accepting Israeli help, were deemed to be equivalent in hindering the government's chances for re-election.

Second, the aid decisions were made in the context of the prior months of rapprochement between Iran and the USA. The two countries had been working to resolve some of their differences at a slow pace and at relatively low diplomatic levels, as described by Armitage (2003). Irrespective of whether Iran needed aid, if the USA had not offered any assistance or if Iran had refused what the USA proffered, that could have caused diplomatic problems. Such an act had the potential for interfering with the détente which both sides had built up and wished to continue pursuing at a low level, out of the public spotlight.

Third, there were other and wider geopolitical considerations influencing the decisions. Iran perhaps did not wish to appear closed to an international presence. That would particularly be the case given the difficulties that the UN nuclear energy inspectors were having in accessing Iranian sites and given that similar intransigence in Iraq was used as an excuse for invading. In parallel, the Bush Presidency was noticing slow progress in Iraq's and Afghanistan's post-war reconstruction. The White House could have felt that another destabilising possibility in the region, which the earthquake disaster might have represented, would not support American goals in the region.

Fourth, both Iran and the USA might have seen incentives in appearing to support humanitarianism. Neither was seen at the time as being a particularly compassionate state, within their regions or around the world. The human-rights records of both states at the time were heavily criticised (AI, 2005; HRW, 2005). Providing post-disaster help, or being at the receiving end of that, yields an opportunity for these governments to show that they support global humanitarian endeavours. That is, they could claim that they will not permit petty politics to intercede with assisting people in need.

24 *Empirical evidence*

This discussion illustrates that many factors beyond the post-earthquake needs were influencing Iran–USA interactions regarding the disaster, the diplomacy, and the disaster diplomacy. The end result was little influence of the earthquake disaster on the attempts by Iran and the USA to forge cautious diplomatic interactions. The earthquake aid did open new doors for communication and had the potential to build trust and goodwill while increasing mutual understanding between the countries. That seemed to have happened between disaster-related workers on the ground. That did not appear to spill over into higher levels or long-term diplomatic successes.

The lack of long-lasting outcomes between Iran and the USA from the Bam earthquake disaster was poignantly illustrated on 22 February 2005. Another earthquake hit southern Iran, killing more than 600 people. The USA offered aid. Iran politely declined the offer. Zarif, still Iran's ambassador to the UN, explained that 'Iran did not refuse the help but said we can handle it domestically'. Domestic capability to deal with a disaster is indeed a factor that would preclude disaster-diplomacy opportunities. In this instance, aid from Algeria, Australia, China, Japan, the United Arab Emirates, and several international organisations was accepted by Iran to deal with the earthquake. That implies that either the disaster could not be dealt with internally or that Iran had political reasons for accepting the aid even if that aid was not needed.

Given that Iran–USA diplomacy had not progressed since the Bam earthquake, Iran presumably felt that political reasons existed for not accepting American assistance. That gesture was reciprocated several months later. In August–September 2005, Iran offered the USA assistance following Hurricane Katrina. The USA displayed little interest in acknowledging that offer (section 4.12).

Despite ample opportunity, Iran–USA disaster diplomacy has yielded few outcomes. Factors and interests other than linking disasters with rapprochement dictated the governments' actions.

4.3 The Philippines from 1990 onwards

The Philippines has had multiple options for disaster diplomacy, none of which lasted for a comparatively long time. Gaillard *et al.* (2009) explored one of the most prominent examples, illustrating the complex links between inter-state and intra-state disaster diplomacy. Gaillard *et al.* (2009) examine the impact of the 1991 Mount Pinatubo eruption on military relations between the USA and the Philippines. The dominant disaster-diplomacy concern was how the Filipino and American governments approached the negotiations for renewing the lease of the two American military facilities in the Philippines. Those negotiations were occurring while the two bases were being damaged by Mount Pinatubo erupting.

The analysis in Gaillard *et al.* (2009) showed that the disaster-related activities due to the Mount Pinatubo eruption had a short-term impact on diplomacy between the USA and the Philippines. This impact was seen in the context of significant connections already existing, through the long-standing American–Filipino military links. In the end, non-disaster factors had a more significant impact on American–Filipino military diplomacy than Mt Pinatubo.

The USA–Philippines bilateral aspect is prominent, making this case study inter-state disaster diplomacy. Additionally, the debates and the reasons for American military involvement in the Philippines has intra-state connections and implications. Many rebel groups around the country identify one aspect of their struggle against Manila as being the dominance of the Americans in governing the Philippines. Many Filipino politicians who supported the closure of the American military bases did so due to their belief that the Philippines was not truly independent with the American military presence. For nation and state building – that is, thinking internally rather than bilaterally or multilaterally – it was important in their view for the bases to go, irrespective of the volcano.

Several purely intra-state examples of disaster diplomacy have also emerged from the Philippines. From November to December 2004, four typhoons struck Quezon province killing over 1,000 people through floods and mudslides in areas with a long-standing guerrilla conflict led by the New People's Army (NPA).

Illegal logging was quickly identified as one of the causes of the devastating slope failures and floods (Gaillard *et al.*, 2007). The Filipino government promptly associated the illegal logging with the NPA. The Filipino opposition blamed the government for not tracking down the loggers and for contributing to the environmental damage. The opposition even suggested the death penalty for the loggers.

An opportunity could have been grasped to tackle the ongoing conflict and illegal logging simultaneously. That could have been linked to long-term disaster-risk reduction, peace, development, and environmental management. Instead, the government and opposition sought to shift blame, to criticise each other, and to inflame ongoing conflicts.

Fanning the NPA conflict on the government side occurred while the parallel conflict with Muslim separatists in the south had cooled down. The situation seemed to be almost as if the government were seeking a conflict somewhere, perhaps to bury the opposition's claims of new evidence for governmental corruption and incompetence. Both sides have media allies promoting their arguments. Instead of disaster diplomacy, this event became politically constructed by the government, the opposition, and the media around the alleged responsibility of nature, illegal loggers, and the NPA (see also Gaillard *et al.*, 2007).

Two other flood disasters plus two volcanic eruptions in the Philippines in 2006 displayed some similar characteristics regarding the political construction of the disaster and the consequent avoidance of disaster diplomacy. In fact, firefights occurred between the NPA and the government's soldiers during relief operations. Conversely, these events also led to proposals for and declarations of ceasefires, sometimes from the government and sometimes from the NPA. Nonetheless, no scope for longer-term peace was suggested and longer-term peace outcomes were not witnessed. Disaster produced short-term, not long-term, diplomatic dividends.

That was a similar result throughout many other incidences of intra-state disaster diplomacy in the Philippines. On 24 November 2007, the Armed Forces of the Philippines declared a unilateral ceasefire in large areas that were about to be affected by a typhoon. The stated that they were suspending offensive military operations against the NPA. That would permit the prioritisation of disaster preparedness and response operations for the imminent landfall of Typhoon Mina (called Typhoon Mitag internationally).

The next day, the Armed Forces of the Philippines expanded this unilateral ceasefire to another area that was in line to be struck by Typhoon Lando (called Typhoon Hagibis internationally). The Communist Party of the Philippines responded to these statements by indicating that unilateral ceasefires and humanitarian assistance are part of their normal operations for places hit by disasters. Both sides agreed, at least on paper for publicity, that military operations were not appropriate in areas about to be or just struck by typhoons. This attitude has not led to conflict reduction between the parties.

Similarly, in September 2009, Typhoon Ondoy (called Typhoon Ketsana internationally) flooded a significant proportion of Manila. On 27 September 2009, guerrillas in the area began a unilateral ceasefire to assist with disaster response. Then, on 15 October 2009, Typhoon Pepeng (called Typhoon Parma internationally) hit northern parts of the Philippines, isolating many areas due to landslides. The guerrillas extended their unilateral ceasefire and explained that they were distributing relief supplies. The death toll from these storms exceeded 700 and led to the short-term disaster diplomacy, but no long-term outcomes.

Taking a geographic perspective of this case study – formulating it as disaster diplomacy in the Philippines – illustrates the overlap between inter-state and intra-state disaster diplomacy. Sovereign states are clearly delineated entities, but inter-state and intra-state case studies involving the Philippines showed few differences. Sovereign entities being one or more of the disaster diplomacy parties might not be a defining feature of a case study.

4.4 Southern Africa 1991–93

Holloway (2000) identifies and examines the drought in southern Africa from 1991 to 1993 as a possible disaster-diplomacy case study. At the time of the drought, significant political changes were ongoing across several countries in the region including the imminent end of Apartheid in South Africa coupled with the end of violent conflicts in Namibia and Mozambique. A drought hit that had the potential to result in a widespread disaster.

The countries of Southern Africa initiated, coordinated, and distributed a large food import from international donations so that the drought emergency did not become a drought disaster. Holloway (2000) detailed the interactions and influences amongst the political and humanitarian processes, concluding that 'while diplomatic dividends can indeed flow from disaster relief efforts, in this instance, joint cooperation was only possible once potential military, economic, and other forms of regional confrontation that dominated the 1980s had been controlled' (Holloway, 2000: 273).

Holloway's (2000) analysis provides further aspects to the long-standing literature on the connections amongst politics, food, agriculture, and drought. Many drought disasters demonstrate how food is widely available at the local level, yet people become malnourished, even starve, because the food is unaffordable to them or is not made available to them. Food costs are determined by political choices on how to set up, regulate, and monitor the food market – locally, regionally, and nationally.

In Southern Africa in 1991–93, the vulnerability of the agricultural system to rainfall variations provided an opportunity for new political cooperation in the region to be solidified. Political choices on how to set up, regulate, and monitor the food markets – locally, regionally, and nationally – were used to counter the effects of poor rainfall. That interaction prevented rather yielded a disaster. But the success regarding the drought-disaster prevention could not have happened without the pre-established political basis for cooperation. That is, diplomacy permitted disaster prevention rather than the potential drought disaster creating drought diplomacy.

4.5 North Korea from 1995 onwards

Since the Korean War ended in 1953, the outside world has had limited access to North Korea. Many disaster-related activities have been undertaken in each Korea and disasters have affected the peninsula. Few diplomatic outcomes were seen for the initial post-war decades, suggesting negligible impact of disaster diplomacy. For example, severe storms and storm surges affecting the Korean peninsula's west coast (Kim *et al.*, 1998) produced few connections or diplomatic results. Instead, the conflict between the Koreas continued in these post-war decades.

A famine in North Korea induced by a series of floods and droughts, but almost definitely caused by agricultural and economic mismanagement, started around 1995. The country was severely disrupted. Death estimates exceeded over a million people. A political minefield emerged regarding international relief operations. Concerns included accountability for aid given, North Korea's re-engagement with the rest of the world (especially outside of China), and the connections between the aid and other diplomacy (Bennett, 1999).

As one example of the increasing connections with North Korea over the years since the disasters started, on 7 March 2000, Japan delivered food aid to North Korea within the context of an agreement for starting diplomatic talks. The ostensibly new era in Korean relations was highlighted on 15 June 2000 when South Korean President Kim Dae-jung visited Pyongyang, including meetings with North Korean President Kim Jong-il. On 18 September 2000, South Korea announced that food aid would be sent to North Korea. On 13 October 2000, Kim Dae-jung was awarded the Nobel Peace Prize for his diplomatic efforts with North Korea.

How much did the floods, droughts, and famine contribute to the North Korea's emergence? When elected South Korea's President in December 1997, Kim Dae-jung had promised a new era of relations with North Korea. He delivered on that pledge. The successes might have been due to North Korea's need for aid even if his reconciliation interests were not influenced by the humanitarian crisis. The large scale of the disaster might be significant, in contrast to the storms and storm surges amongst other smaller-scale disasters throughout the Korean Peninsula. Wider geopolitical contexts, in particular the end of the Cold War, were also influential.

As the aid and diplomacy links progressed, and then failed to progress with North Korea giving few concessions, it became clearer that North Korea was playing games. Pyongyang's response to Kim Dae-jung's election involved pushing the tolerance of those supporting reconciliation. The North was involved in incursions into the South along with missile testing. Pyongyang issued a harsh response to Kim Dae-jung being awarded the Nobel Peace Prize.

The election of George W. Bush to the US Presidency in 2000 did not help due to the foreign-policy approach adopted, as discussed under Hurricane Katrina (section 4.12). Following the terrorism against the USA of 11 September 2001, in January 2002 the Bush administration labelled North Korea as being part of the 'Axis of Evil' (Bush, 2002) for allegedly supporting terrorism.

The ensuing years saw games continue amongst North Korea, South Korea, Japan, the USA, and other countries involved in the negotiations. Several agreements were signed, envoys were exchanged, and meetings were held, but talks went through multiple on-again, off-again phases. That was complicated by North Korea's ambiguity regarding halting its nuclear programme, various military exercises by the different sides, defections, and food shortages in North Korea in parallel with a lack of donations from the international community. On 26 March 2010, North Korea sunk a South Korean navy ship, killing 46 South Koreans and significantly ramping up tensions in the region.

Some minor disaster diplomacy related incidents occurred throughout this time period. In June 2002, North Korea proactively informed South Korea about plans to control water levels at a dam which South Korea had suggested could break, leading to cross-border flooding. North Korea's initiative was welcomed, but North Korea also blasted as propaganda the South Korean concerns about a possible dam break.

On 22 April 2004, two fuel trains collided in Ryongchon, just north of Pyongyang. The resulting explosion reportedly killed approximately 3,000 people, although the exact death toll could not be known due to North Korea blocking information and access. Casualties were shipped to China for treatment and the world offered aid. That included South Korea and the USA, the latter of whose aid went through the UN and Red Cross to reach North Korea.

North Korea initially refused to accept overland aid from South Korea, so one ship arrived on 29 April 2004. The next day, the first humanitarian flight between the two Koreas landed in the North. On 7 May 2004, aid from South Korea finally crossed into North Korea by truck as the two countries agreed to high-level defence talks, which occurred later that month. In June 2004, North Korean and South Korean navies established radio contact for the first time as part of the recent agreements. Border propaganda started being dismantled on both sides and new high-level talks were agreed.

Much of the initial diplomatic activity was spurred on by the explosion disaster, in that negotiations directly linked the aid with progress in reconciliation talks. A significant factor in permitting that to occur was the prior years of interaction over food aid and linked diplomatic talks. A basis had been established for contacts from North Korea to the world while precedents had been established regarding dealing with disaster and diplomacy issues simultaneously. Over subsequent months, the regular diplomatic games recommenced. Links to post-explosion assistance faded with time.

Other possibilities for North Korea disaster diplomacy relate to health and disease (see also section 4.16). Hotez (2004) suggested many ways for pursuing disease diplomacy for North Korea. That included reviving the USA's vaccine diplomacy programme which was distributing vaccines as part of foreign policy. Another suggestion was dealing with the malnutrition and disease resurgence that appeared to be continuing due to the famines. Little substantive measures resulted. One of the few examples was South Korea sending swine flu medicine to North Korea on 18 December 2009 which was reported as being the first aid sent from South Korea's government to the North in nearly two years.

This case study illustrates the complexities of disaster diplomacy. The links are intricate amongst all the aid and diplomacy issues that were placed on the negotiating table. Analysing these links and intricacies is challenging. It might not be feasible to determine how much influence each disaster, and the various disasters in combination, had on the diplomacy, compared to other factors such as the end of the Cold War.

For North Korea, food aid was used to force diplomatic discussions as part of a complex series of delicate negotiations. When aid was less urgent, North Korea became more reticent regarding the diplomacy.

Having established connections amongst people as well as precedents, each series of negotiations had the potential to become less challenging regarding the disaster aid. Confounding factors from the South Korea President's political mandate to the White House's foreign policy to North Korea's own actions influence Pyongyang's responses and willingness to move forward with the international community – and vice versa.

4.6 Cuba–USA from 1998 onwards

In 1959, Fidel Castro overthrew a USA-backed leader in Cuba. In the months that followed, Cuba–US relations deteriorated until by 1960, Cuba was firmly aligned with the USSR. The end of the Cold War did not stop the animosity. Instead, Cuba and the USA continued to be at loggerheads, with the USA maintaining a trade embargo against Cuba and – backed up by international NGOs (e.g. AI, 2010; HRW, 2009) – criticising Cuba for human-rights violations. Meanwhile, Fidel Castro continued to depict the USA as an enemy of Cuba.

Glantz (2000) discusses climate-related challenges and opportunities affecting Cuba–USA relations. His paper covers all disaster-related activities. Examples are scientific and technical collaboration for researching longer-term climatic trends, real-time monitoring of and data exchange for tropical cyclones, and the forecasting and weather impacts of the ENSO (El Niño-Southern Oscillation) cycle, covering El Niño and La Niña.

The cooperation detailed by Glantz (2000) occurred mainly through individual contacts working one-on-one at scientific and technical levels. Higher-level connections that might be publicly visible or that involve more political-level decisions were discouraged by the lack of cooperation between the two countries. These higher-level connections rarely occurred. Glantz's (2000) analysis is that climate-related activities create and maintain links between Cuba and the USA, yet the political differences are not being influenced by those links. In contrast, little political incentive exists for either government to collaborate with the other, whereas scientists and technicians have enjoyed and gained from the mutual help.

A poignant example used by Glantz (2000) is the drought of 1998 in Cuba. It was one of the worst droughts to affect Cuba when Fidel Castro was in power. Cuba's government requested international food assistance through the UN but 'would not accept any aid that obviously came from the US, arguing that food aid was needed in the first place as a result of the US embargo and not because of the drought alone' (Glantz, 2000: 242). Glantz's arguments and conclusions were tested, and corroborated, soon after his paper was published, through several disasters affecting both Cuba and the USA over subsequent years.

On 11 September 2001, terrorists hijacked four American passenger jets over the northeastern USA. Three were deliberately crashed into buildings, one for each tower of the World Trade Center in New York City, leading to the collapse of those structures, and one for the Pentagon in Washington, DC. The fourth plane never reached its target, likely in Washington, DC, because the passengers and crew tried to retake control of the aircraft from the terrorists. The plane crashed during that process. Overall, nearly 3,000 people were killed. Cuba, through Fidel Castro, expressed condolences to the USA. Castro also used the opportunity of his condolence speech to attack the politics of the USA's government.

Two months later, in early November 2001, Hurricane Michelle became the most powerful hurricane to have crossed Cuba since Castro took control of the country. Casualties were minimised due to prior disaster risk reduction measures and advanced emergency management procedures, yet the country experienced extensive damage and Cuba called for international assistance.

The USA offered disaster aid on 7 November 2001. Cuba refused to take the assistance as a donation, but offered to pay for the supplies with the stipulation that Cuban ships transported the goods from the USA. The USA's government did not support this counter-offer. After various counter-offers and negotiations, an agreement was eventually reached. The American food aid was purchased by Cuba but arrived in Havana on 16 December 2001 delivered by an American ship. The final agreement was linked to Cuba–USA trade negotiations that had started in 2000.

The 2005 Atlantic hurricane season provided three more examples of failed disaster diplomacy between Cuba and the USA (Kelman, 2006a). Sixteen people died in Cuba in July 2005 from Hurricane Dennis. The USA's government offered assistance, but the Cuban government politely declined, thanking the Americans for the offer, but preferring to accept aid from Venezuela instead. Several weeks later, the USA returned the favour by refusing Cuba's offer of assistance after Hurricane Katrina killed over 1,800 people across the south-eastern USA (section 4.12). Then, in October 2005, Hurricane Wilma led to hundreds of rescues and damage across Havana. An assessment team was offered by the USA's government which the Cuban government agreed to. Then, Castro changed the terms of reference for the team, which meant that the USA withdrew the offer of assistance.

Despite all these opportunities for Cuba and the USA to try to come together over disaster-related activities, all openings with Fidel Castro leading Cuba were swiftly closed by active efforts to avoid disaster diplomacy from either Cuba, the USA, or both. Ultimately, it appears that rapprochement was not desired. Fidel Castro gained a power base and solidified his power by having the USA as an enemy. Even in the post-Fidel Castro era, cautious attempts at increased connections between the two countries have not been related to disasters and have been hindered by the history of animosity.

On the American side, anti-Castro Cuban-Americans have long had a disproportionate influence on national politics. That is especially the case given that Florida – a focus for the Cuban-American community – proved to be a key battleground for the US Presidential primaries and elections in 2000, 2004, and 2008. Consequently, any movement in Washington, DC towards reconciliation with Fidel Castro's Cuba came at a high political price.

Little incentive has so far existed in either country for disaster diplomacy or for rapprochement more generally. Without Fidel Castro at the helm and perhaps especially after his death, the two countries might slowly come together, with the possibility of a major disaster providing a spark that speeds up the reconciliation. Additionally, the scientific and technical cooperation detailed by Glantz (2000) provides a useful base for inter-governmental connections on disaster-related activities.

4.7 Greece–Turkey from 1999 onwards

The history of disputes between the European neighbours Greece and Turkey dates back centuries. The two modern countries attained their identities and sovereignty largely due to conflict with each other. From 1821–29, the Greek War of Independence was a battle against the Ottoman Empire, which became Turkey after World War I. As Turkey established itself as a post-empire sovereign state, it defeated Greece's army during the Turkish War of Independence 1919–23.

Comparative peace marked Greek–Turkish relations from the late 1920s until the mid-1950s. The 'Greco-Turkish Treaty of Friendship' was signed in 1930 followed by the 1934 'Balkan Pact'. Greece was occupied by the Axis powers during World War II, while Turkey remained neutral until near the war's end when it joined the Allies. Greece and Turkey sent soldiers to fight in the Korean War and joined NATO in 1952. In the mid-1950s, Greek–Turkish relations deteriorated, mainly due to disagreements over Cyprus, and remained frosty and prone to conflict into the 1990s.

On 17 August 1999, over 17,000 people died around north-western Turkey when an earthquake struck. The Greek government along with Greeks responded with an outpouring of public sympathy, donations of money and aid supplies, and rescue teams who pulled Turks from collapsed buildings.

Three weeks later, on 7 September 1999, an earthquake disaster hit the Athens area, killing over 140 people. Post-disaster work in Turkey was still continuing, but the Turkish government and people responded as the Greeks had done. Initiatives came from politicians, rescue teams who pulled Greeks from collapsed buildings, and Turkish people trying to reciprocate what the Greeks had given them. Meanwhile, five other earthquakes hit Turkey from 31 August 1999 causing fatalities until 12 November 1999 when a sixth earthquake in Turkey killed at least 894 people.

From the media and public perspective, a rapid thaw in Greek–Turkish relations followed the earthquakes. Petropoulos (2001), for example, details changes in Greek–Turkish interactions in the form of bilateral agreements (see also Ganapti et al., 2010), tourism, investments and trade, airspace violations, property rights of Greeks in Istanbul (see also Mavrogenis, 2009), and perceptions of each other.

Ker-Lindsay (2000, 2007) explores the influence on Greek–Turkish relations of the earthquake disasters and the mutual aid response. He showed that the rapprochement between the two countries had started before the earthquakes, months earlier in 1999. Those efforts were focused on the two countries' matching concerns over NATO's military action in Serbia and Kosovo. Both countries felt that the situation could yield significant destabilisation concerns for the region.

Empirical evidence 33

The earthquakes, according to Ker-Lindsay (2000, 2007), pushed the Greece–Turkey governmental collaboration into the spotlight. That forced the negotiations to proceed at a much more rapid pace than the politicians and diplomats would have desired. The consequence was many more results than were immediately expected or predicted so soon from the pre-earthquake interactions.

The earthquake disasters influenced the diplomacy, but were not the fundamental cause of it. Ker-Lindsay (2000, 2007) further explains that the rapprochement process might have experienced setbacks, and had the potential for collapse, because such difficult diplomacy, attempting to overcome decades of stereotypes and conflicts, was forced into a public forum. That not only raised expectations regarding rapid results, but also provided open targets for critics trying to interfere with or derail the process.

Mavrogenis (2009) pushes the desire for collaboration between Greece and Turkey back further. He suggests that a major change in attitude in the upper echelons of Greek power occurred at the end of 1995 and at the beginning of 1996. Militaristic posturing between the two countries over some islets of disputed sovereignty nearly led to a full-scale war. Mavrogenis (2009) argues that Greek power brokers realised at that point that violent disputes with Turkey were not an appropriate pathway to select, so they started laying the groundwork to reduce enmity with Turkey.

Meanwhile, Mavrogenis (2009) describes, the Europeanisation process for both countries was a strong incentive to repair relations (see also, for example, Economides, 2005). With Turkey wanting to be in the European Union and with Greece moving towards adopting the euro as its currency, both countries had incentives to become friendlier. The final factor suggested by Mavrogenis (2009) is that, when the first earthquake hit Turkey, Greeks were able to overcome their perception as being an underdog in the Greece–Turkey conflict. Turkey needed help which, according to perceptions, meant that Greece did not have to feel so threatened and overpowered, as they had in the past. All these factors in parallel, in addition to the governmental-level collaboration due to Kosovo, led the earthquake disasters to be an ideal opportunity to push forward rapprochement. That opportunity was grasped.

Ganapti *et al.* (2010) look specifically at Greek–Turkish disaster-related cooperation and how that was affected by the post-1999 disaster diplomacy circumstances. They examined governmental and non-governmental levels of disaster-related collaboration over the decade following the 1999 earthquake disasters. Their evidence is drawn on the peer-reviewed literature, government documents especially bilateral agreements, media analyses, and material from non-governmental organisations. The framing is deliberately a disaster diplomacy lens, so disaster diplomacy relevant material is highlighted.

Based on that material and analysis, Ganapti *et al.* (2010) suggest three conditions under which disasters can lead to and support long-term disaster-related cooperation amongst states that have been in conflict. First, reciprocal gestures of mutual assistance, such as Greece providing aid to Turkey in

August 1999 followed by Turkey providing aid to Greece in September 1999. Second, acceptance that being neighbours means helping each other in times of need. Third, a wider context of connections amongst the enemies would help in founding and maintaining long-term cooperation mechanisms. In the case of Greece and Turkey, that mechanism was the broader rapprochement context; that is, not just disaster-related activities.

Currently, Greece and Turkey are not considered to be adversaries, either internally or externally (for example, Yalçinkaya, 2003). They are even seen to be working to resolve the difficult issue of Cyprus being divided. Disaster-related cooperation amongst other collaborations are seen to be a normal part of bilateral relations. Although the 1999 earthquakes are seen to have influenced the rapprochement, they are not seen as the initiator or the basis for ongoing collaborations.

4.8 Eritrea–Ethiopia 2000–02

In 2000, an opportunity arose for disaster diplomacy between Eritrea and Ethiopia (Kelman, 2006a). Those two countries were embroiled in a border war that began in 1998. In late 1999, it became clear that the region was being affected by a severe drought. Ethiopia soon experienced its worst food crisis in 15 years (Maxwell, 2002). April 2000 saw eight million Ethiopians facing severe food shortages while Eritrea needed food aid for 211,000 people (FAO, 2000; Lortan, 2000).

While formulating the aid response, humanitarian agencies asked Eritrea to allow food aid to be shipped to and unloaded at Eritrean ports, followed by transporting it overland to Ethiopia which, being landlocked, has no ports of its own. Eritrea agreed in April 2000, but this offer was rejected by Ethiopia. Soon after, May 2000 saw some of the war's worst fighting.

In November 2002, the conflict had officially ended, but the border had not yet been entirely defined, as it was still undergoing negotiation and arbitration. Another drought hit, leaving 14 million people in Ethiopia and 1.4 million people in Eritrea requiring food-related assistance. Again, Eritrea made an offer to ship food from Eritrean ports to Ethiopia. Again, Ethiopia refused to accept the offer.

Both times, Ethiopia gave a list of excuses for being unwilling to use Eritrean ports. Numerous allegations were made about Eritrea's behaviour such as Eritrea stealing some of the aid that passes through its territory. Other reasons given were Ethiopia's claims that Eritrea's offer was not serious, but was for public relations only, and that Eritrea made the offer in order to gain the much-needed business of ships docking and aid supplies being unloaded. Ethiopia further stated that non-Eritrean ports were more appropriate as off-loading points and that Ethiopia was not lacking access to ports, but actually needed increased food delivered through the non-Eritrean ports that were already being used.

The attitudes, decisions, and words of both countries seemed to have been affected by a perception of disaster diplomacy being a spectre or a way to gain political advantage over the enemy. By appearing to be compassionate towards its enemy of Ethiopia, Eritrea could have been hoping for credit from the international community which might have translated into international support for Eritrea's territorial claims. Additionally, it was an opportunity to demonstrate Ethiopia's dependence on Eritrea.

Along similar lines, Ethiopia needed to show that the country did not have to rely on Eritrea. There could have been an added incentive in trying to stop any food aid, legitimate or siphoned off, from reaching Eritrea. Another possibility for Ethiopia is that reconciliation or expeditious aid delivery was less important than grasping opportunities to accuse Eritrea of misconduct, in the form of thieving food aid or deliberately aiming to divert aid from the most efficient (non-Eritrean) routes.

Both sides also had a propaganda opportunity that could be used for their own people. Ethiopia could show Eritrea to be heartless in interfering with a 'working' aid delivery system, thereby justifying Ethiopia's need to fight. Eritrea could suggest that there is no point in negotiating with an enemy that starves its own people rather than accepting efficient aid delivery. Such approaches have many precedents. Food aid is used as a weapon of war, both psychologically and physically (McCorkindale, 1994). It is also a human-rights issue, the abuse of which can lead to war-crimes charges (Kent, 2005).

Whatever the excuses given, and irrespective of the veracity or not of the claims and counterclaims, a clear opportunity for disaster diplomacy existed but did not even start at a basic level. Disaster-diplomacy efforts ended up enmeshed within the conflict, with each side using the disaster-diplomacy opportunity to criticise its opponent. Rather than trying to quickly alleviate people's suffering or doing their best to seek diplomatic solutions to the border dispute and to the disaster situation, the conflict was perpetuated by using food aid and possible aid-related diplomatic outcomes as one of the many weapons in the war.

An added dimension is that disaster diplomacy was somewhat of a distraction from wider contexts and could not have addressed all needs. Aside from the discussion regarding the appropriateness or otherwise of Eritrea's ports, other logistical concerns existed regarding transporting food aid through to Ethiopia. Vehicles along the aid routes were commonly attacked by bandits. Personnel on previous convoys carrying aid had been killed because their ethnicity or faction was different than that from the majority in whose territory they were passing through.

Road conditions along the aid delivery routes also hindered transport, from potholes to rains making some locations impassable – an ironic situation given that the aid was for drought relief. Another concern raised was whether or not enough vehicles would be available to transport the needed aid given that the war created a vehicle shortage. Finally, many Ethiopians to whom aid would be delivered had poor access to the areas where vehicles were travelling overland.

36 *Empirical evidence*

The security and access constraints needed to be resolved irrespective of disaster diplomacy's acceptance and implementation by Ethiopia and Eritrea. Disaster diplomacy by itself could do little to deal with those constraints.

More optimistically, if the humanitarian corridor through Eritrea had been implemented, it is possible that the act would have provided a basis for longer-term conflict resolution that would have covered the other concerns. Disaster diplomacy could have illustrated the mutual advantages of cooperating, for droughts, for conflicts, and for the deeper development and sustainability topics facing both countries. Even if routes through other countries were more efficient than delivering aid through Eritrea, the diplomatic gains from using Eritrean routes might have been worthwhile despite the short-term losses for humanitarian assistance.

This reasoning matches political science and international affairs studies explaining that conflict resolution can be supported by taking small cooperative steps that build up trust to tackle larger initiatives (for example, Furlong, 2005). The Egyptian–Israeli peace treaty of 1979 is an example of using earlier confidence-building measures to produce long-term conflict resolution (Landau and Landau, 1997). Whether or not this approach would have worked for Eritrea–Ethiopia disaster diplomacy can be neither proved nor disproved, but it was clear that neither country seemed interested in considering the possibility.

Developments since then leave little optimism regarding disaster diplomacy. A peace deal was reached in June 2000 after the failure of the earlier drought diplomacy. Although violent conflict did not recur, diplomatic conflict did. For example, in April 2003, an international arbitration ruling under the peace deal awarded a border town to Eritrea, but Ethiopia rejected that statement.

Meanwhile, improvements in Ethiopia's harvest did not stop the need for food aid to more than two million Ethiopians at the beginning of 2005. Eritrea also experienced drought in 2005 and over 800,000 people required aid. Despite posturing and military build-ups along the border, a full-scale conflict did not break out again. Instead, as arbitrations by the international tribunal continued through 2009, both sides were generally accepting the outcomes and were seeking to move forward from the conflict.

In these continuing events, drought diplomacy was rarely discussed. Instead, separate efforts seemed to mark the conflict and the famines. Although people's need for food aid was undoubtedly linked to the war, the food aid and the war were seen to be effectively separate issues that were to be solved separately. The Eritrea–Ethiopia case study provides little support for successful disaster diplomacy.

4.9 India–Pakistan in 2001 and 2005

On 26 January 2001, an earthquake shook western India with a death toll exceeding 20,000 people, mainly across Gujarat. Some impacts were felt in

Pakistan, but the disaster was mainly in India. Almost immediately after, Pakistan offered assistance.

One week later, a telephone conversation took place between India's Prime Minister, co-founder and leader of the nationalist BJP, Atal Behari Vajpayee, and Pakistan's military leader and de facto ruler, General Pervez Musharraf (see Abbas, 2005 for background on Musharraf's rise to power and role in Pakistani politics; see Heath, 1999 for some background on the BJP's rise to power). That was the first time that these two leaders had had direct, bilateral contact – at least, in such a public manner. The diplomatic outcome of that contact, of Pakistan's aid offer, and of India's willingness to accept the aid offer included a summit in New Delhi between Vajpayee and Musharraf on 14–16 July 2001.

Less than a month before the summit, on 20 June 2001, Musharraf deposed Pakistan's figurehead President in order to take over the title and position for himself. He was aiming to consolidate his power. There also appeared to be an incentive in gaining leverage for the imminent negotiations with India. Being President Musharraf would be more prestigious than Prime Minister Vajpayee. Motivations influencing Musharraf's political decision-making were different from assisting a neighbour after a disaster or seeking a straightforward route to reconciliation with an enemy.

Despite, or possibly because of, high hopes and intense scrutiny from national and international media, diplomats, and politicians, the July 2001 summit collapsed. A final declaration that would be signed by both leaders was not completed, because they could not agree on one. Some successful initiatives emerged, but the general perception was that the summit had failed even as the two countries' foreign ministers tried to indicate that the summit was a start to, not the end of, dialogue (Dixit, 2002).

The two leaders seemed to have been genuinely seeking, and wishing that they had found, more common ground and ways of moving forward. Vajpayee had agreed to visit Pakistan before the end of 2001, but plans for this visit did not proceed. Through the rest of July and August, Vajpayee and Musharraf exchanged public insults and made few moves to advance reconciliation.

In December 2001, militants carried out several high-profile attacks in India, including infiltrating the Indian parliament in New Delhi. India blamed Pakistan while Pakistan denied any involvement. The countries reduced their links, fortified their military positions along the India–Pakistan border, and ramped up war rhetoric. International diplomats and politicians became worried at the prospect of armed conflict between two nuclear powers that have a history of hostility towards each other and that have fought wars before. They pressured both sides to reduce the tension and to avoid violence.

The situation was further complicated by the Afghanistan situation. Following the terrorism in the USA on 11 September 2001, the USA received UN and NATO mandates to attack Afghanistan because the accused perpetrators of the terrorist attacks were based in and supported by the Taleban government in Kabul. Pakistan had been a Taleban ally, but turned against them after the 11 September 2001 terrorism. The USA-led bombardment of Afghanistan led to the Taleban's overthrow by January 2002, but the Taleban and terrorist leaders were not captured. The international community was working hard to seek stability and reconstruction in the region. That led to a strong incentive to avert any possible India–Pakistan conflict while keeping Pakistan as an ally – especially for fighting terrorism and drug trafficking. The 'earthquake diplomacy' from a year earlier was being superseded by other developments.

March 2002 saw another thaw in India–Pakistan relations, linked to both countries' desire to increase links with the other. The improvement in relations involved restoring many diplomatic ties along with air and bus links, plus – perhaps most importantly from the perspective of diplomacy and international friendliness – cricket test matches. Pakistan made an offer to get rid of its nuclear weapons on the condition that India would do the same, but no agreement resulted.

India–Pakistan relations continued to improve, despite a background of terrorist attacks and moderate earthquakes in both countries. The 26 December 2004 tsunamis (sections 4.10 and 4.11) killed over 10,000 people in India, but had no discernible effect on India–Pakistan connections. One major issue that remains a stumbling block to peace is Kashmir.

Bose (2003) and Schofield (2003) provide background on the wars that India and Pakistan have fought, in addition to the roles and interests of China, with respect to Kashmir. Summarising the current situation, a 'Line of Control' across Kashmir serves as the de facto border between Indian and Pakistan in that region. The two countries' armies face off there. Politically related terrorism and violence related to Kashmir continues on both sides of the Line of Control.

Despite Kashmir and other topics of contention, relations between India and Pakistan were on a slow ascent when, on 8 October 2005, a powerful (moment magnitude 7.6) and shallow (26 km depth) earthquake shook Kashmir, with the epicentre being approximately 105 km NNE of Islamabad. More than 70,000 people were killed in Pakistan including across Pakistan-controlled Kashmir. More than 1,000 people were killed in India including across India-controlled Kashmir. Several deaths occurred in Afghanistan. The immediate response involved many states, including India, offering assistance to Pakistan.

Many commentators suggested the situation as potentially being another form of 'earthquake diplomacy' (for example, Keridis, 2006). In particular, India and Pakistan collaborated to assist the disaster response by lessening restrictions along the Line of Control. On 19 October 2005, telephone links were restored across the Line of Control which enabled family members to contact each other much more easily. Over a nine-day period in November 2005, the governments opened up five locations along the Line of Control for relief supplies to cross. Civilians were soon permitted to cross at one of these checkpoints for the purpose of seeking missing family members.

As these events unfolded, politicians at all levels, the media from local to international reports, and the people in Kashmir and elsewhere in India and Pakistan continually voiced their support for the 'earthquake diplomacy'. The views were that the earthquake could and should signal a new era for Kashmir. Perhaps the dispute could finally be resolved permanently. These hopes were misplaced.

The earthquake disaster brought Kashmir back onto the world stage and assisted in making the Line of Control more porous, thereby tackling the division and hostility induced by that border. To suggest that the earthquake disaster significantly pushed forward the Kashmir peace process neglects wider contexts and prior actions.

In particular, the Kashmir initiatives witnessed after the earthquake were neither a new form of diplomacy linking India and Pakistan nor a new way of dealing with Kashmir. Instead, they represented an acceleration of ongoing India–Pakistan attempts to forge peace and reconciliation, including dealing with Kashmir.

Four days prior to the earthquake, India and Pakistan had agreed to resolve within three months their ongoing dispute regarding troop withdrawals from Kashmir's Siachen Glacier. Months earlier, on 7 April 2005, bus service crossing the Line of Control had started. Despite the threat of terrorist attacks against the bus, the service had been continuing. The openings of the Line of Control after the earthquake disaster were important, but they were not unprecedented and they could not by themselves end Kashmir-related disputes and violence.

Furthermore, despite the disaster and the peace overtures, violence continued in India and Pakistan, including across Kashmir. In the short term (weeks and months) following the earthquake, India-controlled Kashmir's education minister was assassinated (18 October 2005), bomb attacks in New Delhi killed over 50 people (29 October 2005), three people were killed in a bombing of a fast-food outlet in Karachi (15 November 2005), and one of Pakistan's main gas pipelines was damaged in a bomb attack (28 December 2005). In the long term (a time scale of years), Mumbai suffered with over 160 dead in each of train bombings on 11 July 2006 and co-ordinated attacks in late November 2009. Meanwhile, locations around Pakistan have continued to be hit by suicide attacks, sometimes killing dozens.

The widespread nature of the violence, with most attacks involving various non-state groups which might or might not be supported by states, illustrates that the violence in India and Pakistan emanates from concerns wider than Kashmir and results from more than inter-state conflict. Differences in cultural values and religion are often cited as reasons for the attacks, not just an India-versus-Pakistan situation.

Kashmir is not as simple as a conflict between India and Pakistan (for example, Bose, 2003; Mohan, 1992; Thorner, 1949). Those promoting and enacting violence in Kashmir seek different forms of control, or lack of control, by India and Pakistan over different parts of Kashmir. Some groups aim for independence. Drugs feed the conflict and China has interests in the region. Many different levels emerge, including peace in Kashmir, peace in India, peace in Pakistan, peace between India and Pakistan, and peace amongst India, Pakistan, and China. Much more than Islamabad–New Delhi reconciliation is needed, yet that was the focus of both the Gujarat and Kashmir attempts at 'earthquake diplomacy'. Perhaps, that is why they both failed.

In the meantime, many people in, and the governments of, India and Pakistan continue to work hard to reduce tension and to foster peace between the two countries, irrespective of disasters. The April 2010 wedding between a Pakistani cricket star and an Indian tennis star might have had more influence on the population's support for Indian–Pakistani reconciliation than either earthquake.

4.10 26 December 2004 tsunamis: Sri Lanka and Aceh

On 26 December 2004, one of the largest magnitude earthquakes (moment magnitude 9.1) over the previous century struck at a shallow depth (30 km) off the coast of Indonesia. Following the earthquake's localised destruction, tsunami waves dozens of metres high slammed into the coast of an Indonesian island, Aceh. The waves had propagated in all directions from the earthquake's epicentre, eventually crossing the Indian Ocean after several hours. Over 250,000 people were killed in more than a dozen countries including Sri Lanka, with Indonesia experiencing the most devastating effects and the highest mortality.

Due to the presence of thousands of visitors from foreign countries in the inundated areas, many affluent countries were directly affected, including major players in regional and international politics, notably the USA, the European Union, and Australia. Politicians in Canada, New Zealand, Sweden, and the UK were lambasted for their initially lackadaisical response to the disaster because hundreds of those countries' citizens were missing or affected. This criticism and the foreigners affected turned an initially slow response from many donors, partly due to the holiday season in many parts of the world, into one of the largest ever global humanitarian responses up until that time.

The response, recovery, and reconstruction strategies were further challenged by the presence of various levels and forms of conflict in many of the affected areas. Most prominent were internal conflicts in Aceh and Sri Lanka which the international community had long tried to resolve.

In Sri Lanka, the conflict is rooted in a decades-long history evolving from differences between the majority Sinhalese on the island and one of the larger minorities, the Tamils (for example, Chattopadhyaya, 1994; Singer, 1992; Tambiah, 1986; Wilson, 2000). Muslims form another sizeable minority in Sri

Lanka. The Liberation Tigers of Tamil Eelam (LTTE; Tamil Tigers) were founded in 1976 and started launching major attacks in 1983. As the LTTE's power grew, they assimilated or fought many other Tamil groups that were pursuing violent or non-violent means for Tamil rights. The LTTE soon became the main violent group on the Tamil side, epitomising the Tamil side of the violent conflict.

The LTTE also waged international campaigns, seeking sympathisers, funds, and arms through the large international Tamil diaspora. A de facto state was set up in the northern and eastern areas of Sri Lanka under LTTE control, with attacks on the rest of Sri Lanka including the LTTE's own air force and navy. Two of the most audacious attacks were suicide bombers killing (i) former Indian Prime Minister Rajiv Gandhi in 1991 in southern India as he campaigned for re-election and (ii) the Sri Lankan President Ranasinghe Premadasa in 1993.

Prior to the tsunami, at least one disaster-diplomacy related incident had occurred. Following a ceasefire between the LTTE and the Sri Lankan government in February 2002, tensions were rising in 2003 when flooding hit southern Sri Lanka in May, killing at least 250 people. The LTTE organised a relief convoy and donated supplies to assist with the response. No discernible outcomes on the peace process were observed.

Sri Lanka then suffered the second highest mortality from the 2004 tsunami disaster, with 30,000 dead and approximately half a million homeless. Some of the worst-hit areas were Tamil-controlled territory. The rebels responded quickly and effectively, because they knew and controlled the territory, but they did not immediately have access to aid that was channelled through the Sri Lankan government. After months of wrangling, at the end of June 2005, an agreement to share aid was reached. Days later, based on a challenge on constitutional grounds by Sinhalese nationalists that the LTTE are terrorists without legal status, Sri Lanka's Supreme Court blocked the agreement. Sri Lanka's government showed little inclination to try to revive or revise the deal.

During this process, bickering over aid provision and prioritisation led to perceptions and accusations that emergency aid had been unequally distributed and that reconstruction policies were socially, politically, and culturally biased, by failing to factor in different needs of different populations (Enia, 2008, 2009; Le Billon and Waizenegger, 2007; Rajagopalan, 2006; Rajasingham-Senanayake, 2005; Renner and Chafe, 2007; Uyangoda, 2005). Internal disputes amongst different groups of Tamil fighters and amongst different parties within the Sri Lankan government contributed to the lack of desire to reconcile. Parts of the criticisms that various sides in the conflicts levelled at each other were opportunistic, designed to foment, rather than to resolve, conflict.

An unwillingness to cooperate led both sides to make public accusations. The government accused the rebels of not providing safety guarantees for aid delivery and of wishing to siphon off aid. The rebels accused the government of discriminating against Tamils while deliberately withholding aid. As with Eritrea–Ethiopia (section 4.8), both sides used the disaster to perpetuate the same accusations and actions that had marked the pre-tsunami conflict. As the post-tsunami months progressed, the situation worsened.

In August 2005, Sri Lanka's foreign minister and strong LTTE critic was assassinated in his home. A national state of emergency was imposed. Within a week, though, Sri Lanka's government and the LTTE agreed to hold high-level talks. But they could not agree on the logistics, with continued wrangling going into September, further degrading trust.

17 November 2005 witnessed Presidential elections. Mahinda Rajapaksa, promising to take a hard line with the LTTE including a re-negotiation of the ceasefire, won Sri Lanka's Presidency, battering an already fracturing peace process. Violence increased across Sri Lanka in the following months. Despite that, in April 2006, Rajapaksa's party won overwhelmingly in local elections, endorsing his approach to dealing with the LTTE.

Fighting escalated and international observers withdrew, with the military balance slowly shifting in favour of Sri Lanka's government. By mid-2009, fighting was finished in Sri Lanka and most of the LTTE leaders were dead, by suicide or in combat. Rajapaksa was re-elected President overwhelmingly in January 2010, receiving little support in Tamil areas of the country.

Tsunami diplomacy had failed entirely in post-tsunami Sri Lanka. It is difficult to pinpoint whether or not the beginning of the end of the LTTE was the tsunami, was the disinterest of the parties in conflict to use the tsunami disaster as an opportunity to seek peace, or was inevitable irrespective of other factors such as disasters.

Aceh had more scope for disaster diplomacy, with the background to the case study summarised here from Aspinall and Berger (2001), He and Reid (2004), Mardhatillah (2004), Reid (2005, 2006), Sulistiyanto (2001), and van Dijk (1981). As with Sri Lanka, the roots of the conflict go back decades.

Aceh was a successful independent sultanate until the Dutch forcibly took control in the early twentieth century. Pro-independence insurgencies against the Dutch continued until World War II. Indonesia, encompassing Aceh, declared independence in 1945, and independence was formally attained in 1949. Aceh separatist movements continued, embroiled in concerns covering numerous territorial, religious, social, and economic aspects, including control of natural resources.

In 1976, GAM (Gerakan Aceh Merdeka or Free Aceh Movement) was founded and soon became the main leader of Aceh's independence battle. Several attempts to reach peace settlements faltered. In August 2001, Indonesia's government passed a law which gave Aceh increased autonomy over its internal affairs. On 9 December 2002, a Cessation of Hostilities Agreement between GAM and Jakarta was signed.

Five months later, the Cessation of Hostilities Agreement fell apart as the efforts to implement the autonomy law slowly failed. Aceh was declared to be in a state of military emergency in May 2003, which a year later was downgraded to a state of civil emergency. Indonesia elected a new president, Susilo Bambang Yudhoyono, in October 2004 who visited Aceh in November 2004, indicating interest in trying to reach a peaceful settlement. Then, the tsunami struck, killing more than 165,000 people and leaving half a million homeless from Aceh's population of just over four million.

On 23 January 2005, the Indonesian government and GAM announced that they had agreed to resume peace talks in Helsinki. The preliminary discussions ended on 23 February 2005 with optimism. The mediators along with both sides emphasised the progress that had been made and highlighted the desire and intent to reach a lasting settlement. On 15 August, GAM and the Indonesian government signed a Memorandum of Understanding that was heralded as a peace deal for Aceh. Despite frequent, small-scale violations, the deal has held since then, appearing to be a path for permanent peace. From the tsunami's devastation, disaster diplomacy appears to have birthed conflict resolution.

Gaillard et al. (2008) show that the situation was not so simple. Yudhoyono was elected President with Y. Kalla as Vice President. In the months prior to the October elections, Yalla had established initial contact with GAM negotiators. As Vice President, Kalla was given authority to pursue peace talks. On 31 October 2004, a secret deal was signed between the Indonesian government and GAM's military chief for Aceh, as a first step towards a wider peace process.

The deal failed when it was revealed by Acehnese media and when it appeared to benefit government representatives more than people affected by the conflict. This failed deal provided the foundation for further discussions that led to formal, secret negotiations between GAM and the Indonesian government starting on 24 December 2004 – two days before the tsunami catastrophe. These secret discussions appear to have been the basis that led to the 2005 Helsinki peace talks and Memorandum of Understanding.

Enia (2008, 2009), Le Billon and Waizenegger (2007), Morfit (2006), Renner and Chafe (2007), and Schulze (2005) further lay out a series of non-tsunami factors that contributed towards the success of the peace process. Yudhoyono and Kalla were willing to take political risks over Aceh. The Indonesian military backed a peaceful solution. GAM's military forces had been significantly weakened prior to the tsunami, during the Indonesian military's 2003 offensive. Solid international mediators kept the two sides talking. Indonesia's delegation negotiating with GAM was not steeped in the usual, Java-based, bureaucratic culture.

Consequently, numerous pre-tsunami military and political contexts created a basis of conditions conducive to a lasting peace. The tsunami created the international interest and domestic political space that could make successful use of those conditions. That is, a basis for peace in Aceh existed before the tsunami, but the tsunami ensured that this basis did lead to peace rather than succumbing to failure, as had happened before.

44 Empirical evidence

It can never be known whether or not peace would have been achieved without the tsunami. It is clear that the tsunami disaster deeply and positively affected the ongoing reconciliation attempts between Aceh and Djakarta. One major influence was the sudden and extensive international interest in Aceh due to the tsunami coupled with the necessity of opening up Aceh to international assistance.

Gaillard et al. (2008) make an interesting point in that regard. Their field evidence suggests that, prior to the tsunami, only a few international parties were interested in resolving the Aceh conflict. After the tsunami, most international parties involved focused on tsunami aid, without providing much clout or financing for the conflict resolution. For instance, considering the largest post-conflict donors for Aceh, the EU pledged US$20 million for post-conflict reconstruction compared with US$250 million for post-tsunami activities. USAID's figures were US$10 million and US$400 million respectively. As that was happening, those interested in conflict resolution continued their quiet diplomacy to seek peace within the context of the tsunami's devastation.

Nevertheless, separating reconstruction from conflict and separating reconstruction from other disasters is not simple in Aceh or in other case studies. That is especially the case since successful disaster recovery and successful conflict resolution tend to be intertwined and connected to longer-term disaster risk reduction, development, and sustainability processes (for example, Lewis, 1999; Renner and Chafe, 2007; Wisner, 2003).

To a large degree, splitting post-tsunami reconstruction and post-conflict reconstruction was neither attempted nor completed in Aceh. Consequently, post-tsunami reconstruction funds available for Aceh contributed to developing peace. Many, although not all, aspects of reconstruction were linked between the tsunami and the conflict – even to the point that slow, inequitable, and incompetent reconstruction practices were seen for both tsunami-related and conflict-related activities. While some humanitarian workers focused strictly on their tsunami relief mission, others inadvertently linked or were forced to link their tsunami reconstruction activities with rehabilitation and reintegration of combatants, to assist them in becoming ex-combatants.

Elements of disaster diplomacy could not have been avoided in Aceh, even when that was not the expressed aim or wish. Aceh tsunami diplomacy illustrates how a disaster can significantly spur on a peace process that existed already, especially by creating a space where peace is feasible, even though the disaster does not create that peace process. The contrast with Sri Lanka is poignant where pre-existing conditions provided the excuses to use the tsunami to exacerbate the conflict.

4.11 26 December 2004 tsunamis: other locations

In addition to Aceh and Sri Lanka (section 4.10), preliminary post-tsunami political reactions indicated possibilities for several other locations with the potential for tsunami diplomacy. Most of the hopes in these potential case studies faded rapidly, as had been predicted by Kelman (2005). Part of that might be attributed to world attention focusing on Aceh and Sri Lanka regarding tsunami diplomacy.

As discussed in section 4.13, Burma had an internationally isolated regime with intense internal conflicts regarding the country's governance. While tsunami damage did not appear to be extensive, fatalities were reported and it is hard to obtain a complete picture due to the country's restrictions. No change in Burma's governance, disaster response, international access, or information flow was seen after the tsunami. Lack of disaster-related changes were further highlighted in the inadequate response to Cyclone Nargis (section 4.13).

When the tsunami struck, the Maldives was also governed by an oppressive dictatorship. The country comprising low-lying atolls was badly affected by the tsunami, with over 80 killed from a population of nearly 400,000. The tourism and hospitality industry, a predominant livelihood, suffered extensive damage. Many non-tourism livelihoods such as fishing were harmed as well. Government–opposition reconciliation was suggested given that the country's President dropped treason charges against opposition leaders in the tsunami's aftermath.

Little further diplomacy resulted. 2005 saw continual government–opposition conflict. Opposition leaders were later charged again with anti-government offences. Due to political factors not linked to the tsunami, the Maldives' President and Parliament eventually agreed to the country's first multi-party, open elections in 2008. On 29 October 2008, the President conceded defeat to the opposition and he worked for a cooperative and smooth power transition. Democracy came to the Maldives without much role being played by one of the worst disasters to have affected the country recently.

The tsunami struck Somalia, killing dozens. Somalia has been ungoverned as a country since 1991 when a dictatorship collapsed. While some regions within the country have established somewhat stable governments and governance, the country is marked by violence and warlord control. Hopes were expressed that the tsunami's impact might induce the world to try to support the country back to stability. Some minimal international efforts continue to support Somalia's people and to aim for an accepted, democratic, national government. Those efforts have not been linked to the tsunami and the tsunami had little influence on political outcomes or lack thereof.

India was seen as potentially being pivotal on several fronts regarding tsunami diplomacy. Little change resulted, even though parts of south-eastern India along with the Andaman and Nicobar islands were devastated by the tsunami, leaving approximately 10,000 dead. India's initial response was that the country did not want or need any tsunami-related foreign aid.

That statement was seen as particularly snubbing the Americans, a disappointment to those who hoped that India and the USA would become closer as a result of the tsunami. India and the USA are not traditionally seen as allies (Kronstadt, 2004; Sagar, 2004), but trade is a powerful force behind some politicians in both countries seeking closer connections. Even the 11

September 2001 terrorist attacks were seen as bringing India and the USA slightly closer together (Kronstadt, 2004), although the change was temporary. Subsequent political decisions from both countries, mostly regarding foreign policy, annoyed the other and stymied attempts at reconciliation (Sagar, 2004).

US President Bush visited India in 2006, bringing the two countries closer together. The trip was labelled as being important for security and trade, particularly with regards to a deal for the USA to support nuclear energy in India. The trip was not linked to either the 2004 tsunami in India or to the 2001 terrorist attacks in the USA.

India's reluctance to accept tsunami-related foreign aid was not seen as an attempt to avoid assistance from Pakistan. At the time, India–Pakistan disaster diplomacy from the 2001 Gujarat earthquake (section 4.9) was viewed as having petered out. The two countries' relations were assumed to be affected mainly by political influences and political outcomes, rather than being directed by disasters such as the 2001 earthquake or the 2004 tsunami. As discussed in section 4.9, this view was corroborated after the 2005 earthquake.

Positive relations were seen between India and Sri Lanka following the tsunami. Both countries had empathy with the other being affected, plus India immediately offered aid to Sri Lanka. These connections need to be seen in the long term, generally positive diplomatic and political links between both countries. While it shows how neighbours experiencing their own conflicts can help each other post-disaster, it would be a stretch to suggest India–Sri Lanka as a disaster-diplomacy case study.

Finally, with regards to India, at the time of the tsunami disaster (and the 2005 earthquake; see section 4.9), India did not share real-time seismic data. India's government gives one reason as being the need to process such data to ensure its robustness, rather than releasing raw numbers that could be misinterpreted or misused. Another interpretation is that India's reticence to release such data is to avoid scrutiny of its nuclear programme that includes testing nuclear bombs.

Both the 2004 tsunami and the 2005 earthquake brought India under increasing pressure to change its policy and to share real-time seismic data to assist with earthquake and tsunami response. Changes have been happening, including as part of the Indian Ocean tsunami warning system. Full and open access to seismic data from India is still not feasible.

Aside from Aceh (section 4.10), another Indonesia-related tsunami diplomacy possibility suggested in media and government discussions was improving Indonesia–USA relations. The basis was the involvement of the American military in bringing humanitarian relief to tsunami-affected locations in Indonesia, including Aceh. To consider this case study as being possible disaster diplomacy seemed to be almost wishful thinking – attempting to create a 'peace from the ruins' story without much basis, demonstrating the intense false hope that ideas of disaster diplomacy can generate.

Following the tsunami, the American and Indonesian governments both suggested that the American military operation to bring relief supplies to Indonesia would support and possibly improve Indonesian–American relations, a story picked up and amplified by the media. The two countries, though, have usually had relatively close and friendly relations, covering military, economic (including oil), and political connections.

Incidents occasionally cause friction. An example occurred in October 2000 when the new American ambassador in Jakarta decided to take on corruption in the Indonesian government. Irrespective of this, little intimation exists that Indonesia and the USA are in conflict.

USDS (2002) summed up Indonesian–American ties in recent times, covering the then-new Bush foreign policy as well as the pre-Bush-era, in that 'The United States views Indonesia as the cornerstone of regional security in Southeast Asia and a key trade partner. U.S. interests in the region depend on Indonesia's stability and economic growth' (online at the time, but the source no longer exists). It is typical of the USA's government's attitude that trade and security, even after the 11 September 2001 terrorist attacks, are implied as being equivalent and that economic growth is suggested as being as important as political stability. The same American government department, in referring to 2002, commented that 'The [Indonesian] government's human rights record remained poor, and it continued to commit serious abuses ... These abuses were most apparent in Aceh Province' (USDS, 2003: online at the time, but the source no longer exists).

These human-rights concerns were not permitted to interfere with warm cooperation between the two countries. USAID (2002) planned US$124 million of development assistance and economic support to Indonesia for the 2002 fiscal year, further indicating that a request would be made for US$131 million covering similar activities for the 2003 fiscal year. In 2003, the USA was Indonesia's second largest trading partner, with 12.1 per cent of Indonesia's exports going to Americans (CIA, 2005).

Yet the American government felt, or chose to suggest, that the Indonesian people might not be overly friendly towards the USA. Consequently, US Secretary of State Colin Powell considered that, in witnessing American aid to Aceh and other tsunami-affected parts of Indonesia, the people of Indonesia might become friendlier towards the USA and towards Americans. He stated that post-tsunami relief supplies and support from the American military was 'an opportunity to see American generosity, American values in action' (for example, Reynolds, 2005: online).

Powell's view was criticised as being naïve (Kelman, 2005) given that many other countries did not reduce their animosity to the USA or to Americans despite also experiencing 'American generosity, American values in action' through recent post-disaster aid. Examples are Iran (section 4.2) and North Korea (section 4.5). Even if Powell felt that foreign policy should be altered when recipients are donated aid from the USA, given past experience, was it realistic to expect a substantive and lasting shift in public opinion, or a sudden acceleration of Indonesia–USA cooperation, given only the tsunami response without any other factors?

A deep level of anti-American and anti-Western antipathy exists in some small sectors of Indonesian society, leading to numerous terrorist attacks. Australians are often principal targets. Instances targeting Western icons or citizens in Indonesia in the past decade include bombing tourist sites in Bali on 12 October 2002 and 1 October 2005 (attacks also designed to hit Bali's mainly non-Muslim local population); bombing Australia's embassy in Jakarta on 9 September 2004; and bombing two Jakarta hotels favoured by Westerners on 17 July 2009.

Meanwhile, the vast majority of Indonesians would never resort to violence and are harmed by these terrorist attacks as casualties as well as through loss of livelihoods. They are much more concerned with day-to-day living, having limited concerns regarding international geopolitics. While post-disaster aid is always gratefully received, from the USA or other places, it would not be seen as an anti-terrorism measure since most Indonesians have little interest in terrorism in the first place.

Despite these multiple instances of lack of immediate success of tsunami diplomacy from the 26 December 2004 disaster, one form of international cooperation did result. Due to the disaster, the Indian Ocean tsunami-warning system was designed and made operational on 28 June 2006, just 18 months too late for those who died. As with the Pacific tsunami-warning system, the cooperation is almost exclusively at a scientific and technical level. That is a form of disaster diplomacy in itself; for example, see the Cuba–USA case study in section 4.6. A useful study remains to be completed to explore how and why that scientific and technical cooperation has or has not led to other forms of cooperation, especially at higher political levels.

4.12 Hurricane Katrina in 2005

In January 2001, George W. Bush was inaugurated as President of the USA. That followed a closely fought election in 2000. It took weeks to decide the election's outcome and the election continues to be debated and disputed, over allegations of fraud and abuse of power in accepting and rejecting ballots during the count (Posner, 2001). The Bush White House immediately pursued a foreign policy disliked by other states and generating various forms of conflict worldwide (see Camroux and Okfen, 2004; Dunn, 2003; Dunne, 2003; George, 2005; Hurst, 2005). The policy was based on three core elements:

- *Isolationism*: The USA would be less involved in world affairs; for example, having a low profile in African conflict resolution (see also Schraeder, 2001).
- *Hegemony*: American foreign policy would spread the values and ideas which the American government deems are important. These American values and ideas were usually referred to, with limited clarification, as

words or concepts such as freedom, democracy, prosperity, and security (see also Blum, 2003 and Chen, 2003).
- *Unilateralism*: The USA would make international affairs decisions with minimal consultation with other states (see also Lake, 2006 and Skidmore, 2005).

These elements were reinterpreted but reinforced following the 11 September 2001 terrorist attacks in the north-eastern USA (Beeson and Higgott, 2005; Leffler, 2005; Parmer, 2005; Skidmore, 2005; US Government, 2002). Despite the international outpouring of sympathy for the USA suffering from heinous terrorism, any goodwill was quickly squandered by the White House's inability to engage with the world – another abject failure of disaster diplomacy.

Unilateralism was rendered 'a new multilateralism' meaning that 'the United States unilaterally defines a global agenda that everyone who agrees with is welcome to accomplish alongside the United States' (Camroux and Okfen, 2004: 174; see also Chen, 2003 and Dunne, 2003). This approach could be termed 'a new unilateralism' too. The American government built a coalition of states supporting their interests, especially military action. Any state which declined to join would no longer be favoured by the American government. Isolationism was somewhat weakened through creating coalitions, yet somewhat strengthened by isolating the USA from states which disagreed with its foreign policy. Hegemony was emphasised because the terrorism was seen as attacking American values.

US Government (2002) describes these aspects of the new foreign policy. A poignant example is 'The U.S. national security strategy will be based on a distinctly American internationalism that reflects the union of our values and our national interests' (p. 1). The document then continues unambiguously, 'While the United States will constantly strive to enlist the support of the international community, we will not hesitate to act alone, if necessary, to exercise our right of self-defense by acting preemptively against such terrorists' (p. 6). A third example is explaining that 'In exercising our leadership, we will respect the values, judgment, and interests of our friends and partners. Still, we will be prepared to act apart when our interests and unique responsibilities require' (p. 31).

Strong international support, including through the UN and NATO, existed for the American-led war in Afghanistan later in 2001 as a response to the 2001 terrorist attacks. Any goodwill for the Bush administration fell apart when the American government put together a coalition of countries to attack Iraq in 2003 without explicit or affirmed support from international institutions. Bush won another narrow and disputed victory in the Presidential elections of 2004.

At the end of August 2005, a powerful hurricane named Katrina made landfall along the Gulf Coast of the USA, killing approximately 2,000 people and revealing significant inadequacies in the federal, state, and local preparation for and response to these events (DHS, 2006; Dunn, 2006; GAO,

2006; US Government, 2006; US House of Representatives, 2006). The difficulties which the world's richest and most powerful nation experienced in effectively managing an event which had been predicted (for example, Fischetti, 2001) and planned for (for example, OEP, 2000) surprised Americans along with people and governments around the world.

The predominant international response was generosity, involving donations and offers of aid, including numerous states with varying degrees of conflict with the USA (Table 4.2). Venezuela's offer, for example, came despite the American religious leader (and one of the 1988 losing candidates for the Republican-party nomination for US President) Pat Robertson suggesting in August 2005 that Venezuela's President should be assassinated. The American government's initial reaction to the offers of assistance was to try to avoid accepting external help. Initially, the American government did not even acknowledge the offers from some states, such as Cuba, Iran, and Venezuela. The ensuing days led to a series of confusing and often contradictory briefings from the State Department regarding the foreign aid (USDS, 2005a–f).

USDS (2005f) finally acknowledged the immense outpouring of sympathy and support from around the world, but noted only who had pledged assistance without commenting on whether or not the offers were needed or were accepted. In USDS (2005c, 2005d), several direct questions regarding the diplomatic implications of accepting aid – especially from Cuba, Iran, and Venezuela – were posed by the media to the US State Department. Those questions were not answered. In USDS (2005e), the US State Department mentioned that no offers of assistance had been refused and then declined to answer a question regarding Iran's offer of aid.

The confusion in dealing with unusual situations – being overwhelmed by a domestic disaster and not expecting or being used to offers of international assistance – was a significant factor in the American government's uncoordinated response. The perpetuation of the American government's isolationist and unilateralist foreign policy was effected mainly by default.

Kelman (2007) also described that inexperience in dealing with the new situation posed by Katrina led to inertia and inflexibility in foreign-policy responses. That argument might be flawed considering the USA's long history of dealing with disasters of varying sizes at the federal level. The Federal Emergency Management Agency (FEMA) was created exactly for that purpose. Yet after the 11 September 2001 terrorist attacks, FEMA was restructured and placed within the Department of Homeland Security rather than being its own agency. Many key personnel with successful operational experience left, weakening the agency in the eyes of its own employees (Menzel, 2006; Perrow, 2005).

Since January 2003, FEMA had been led by Michael Brown whose only disaster-related experience prior to joining FEMA in 2001 had been more than twenty years previously. Brown's inexperience with crises and inability in leading FEMA have been cited as contributing to the overall amateurish response by the USA's government to Katrina (Sobel and Leeson, 2006).

Table 4.2 Examples of Hurricane Katrina aid offered by states in conflict with the USA

State	Conflict context	Aid offered
China	China–USA relations were marked by diplomatic conflict over Taiwan, Hong Kong, Tibet, and Iraq countered by some collaboration on trade and North Korea	Five million dollars of aid, rescue workers, and medical experts
Cuba	The USA opposed Fidel Castro's regime and had maintained trade sanctions against Cuba	More than 1,000 doctors and several tonnes of medical supplies
France	In the run-up to the 2003 Iraq war, France had threatened to veto any new United Nations Security Council resolution authorising the use of force against Iraq	Tents, generators, and water-purifying plants amongst other materials, including communications and water-treatment facilities from French organisations and companies
Germany	Germany–USA relations were frosty due to Germany's opposition to the Iraq war	25 tonnes of food, emergency shelter, and services for medicine, transportation, water treatment, and search and rescue
India	India–USA relations have not traditionally been overly friendly. The 2001 terrorist attacks briefly brought the countries closer together	$5 million, medicine, a medical team, and water-purification systems
Iran	Since a 1979 revolution overthrew the American-backed leader, Iran and the USA have been on unfriendly terms	20 million barrels of crude oil and assistance for the rescue efforts
Mexico	Diplomatic conflicts occur over border control related to immigration, drugs, pollution, and terrorism	$1 million, water, food, medical supplies, vehicles, and equipment, delivered to the Gulf Coast by the Mexican army
Russia	Since the collapse of the USSR, Russia has tried to resist American hegemony. Political and diplomatic conflicts have emerged between Russia and the USA, particularly over the Middle East and the 2003 Iraq war	Medical supplies, rescue teams, helicopters, food, tents, blankets, drinking water, and portable electricity generators
Spain	In 2004, Spain elected a government opposed to the Iraq war, which immediately withdrew Spain's troops from Iraq despite intense American objections	Oil, food, batteries, medicine, and a Red Cross delegation
Venezuela	The views of Venezuela's President Hugo Chavez are disliked by the American government. The USA complains about drug and human trafficking from Venezuela while Chavez accuses the USA of supporting his opposition	Food, oil, water and aid workers plus soldiers to help tackle reported looting in New Orleans

Source: Kelman, 2007

52 *Empirical evidence*

Despite the domestic-response troubles, the USA's government's foreign-policy position on aid evolved and softened to some degree. Part of the evolution was spinning the situation to appear more open than the government actually was. For example, USDS (2005b) made it clear that international aid was needed, that the offers were appreciated, and that aid was being accepted. Yet Cuba's, Iran's, and Venezuela's offers were not acknowledged until several days after the offers were made. Events on the ground were also overtaking the government's official response. For instance, on 1 September 2005, the United States Customs and Border Protection 'facilitated an expedited entry of relief supplies from foreign countries' (US House of Representatives, 2006).

Although undercurrents of hegemony towards other states could be interpreted in the USA's government's reactions (USDS, 2005a–f), the American values espoused by the White House's foreign policy was a relatively small factor in the USA's government's response. Aggressiveness was never apparent in the USA's position on post-Katrina international aid. Instead, the USA's government appeared grateful that the world was willing to assist the USA, but incompetent in trying to manage that aid.

Rather than a deliberate or calculated snub, the USA's government seemed to be at a loss regarding how to deal with other states during domestic disasters – especially dealing with the ostensibly mystifying responses from enemy countries in offering assistance. That reaction was despite the previous American offers of post-disaster aid to Cuba (section 4.6) and Iran (section 4.2). The distraction caused by the hurricane's devastation, the uncertainties in how to deal with the disaster at a domestic level, and the engrained approach of hegemony, isolationism, and unilateralism on the world stage served to overwhelm the rapidly changing international situation and the implications of the aid offers.

This lack of adaptation of the USA's government's foreign policy, deliberate or otherwise, effectively precluded any significant disaster-diplomacy outcomes through the USA being a recipient of aid. That contrasts with the expectation implied by Powell after the 2004 tsunamis that aid from the USA would make others feel more friendly towards the USA (section 4.11).

At the time of Katrina, however, Powell was no longer the US Secretary of State. Demanding consistency between the two events might be unfair. Nonetheless, that might have been the desired result from the USA's perspective, either through planning or from lack of planning leading to inertia. The consequence is that Hurricane Katrina disaster diplomacy did not last in the long term and did not influence the USA's relations with the world.

4.13 Two May 2008 disasters

May 2008 brought a useful contrast of disaster-diplomacy case studies based on specific disasters. On 1 May 2008, warnings regarding Cyclone Nargis, a major Bay of Bengal storm, were issued, including for Burma's (Myanmar's) coastal areas. On 2–3 May, the cyclone struck Burma, apparently with little preparation having been completed. The Burmese government's official death toll was approximately 22,000, but the actual death toll might have exceeded 100,000.

Burma's government was a military dictatorship, ostracised by most of the world for refusing to budge over human rights and for further isolating itself by keeping the country extremely closed. The leader of the opposition Aung San Suu Kyi was under house arrest for promoting democracy, with her house arrest scheduled to end on 27 May 2008. Additionally, the government was gearing up for a referendum on its proposed constitution for 10 May, just one week after the cyclone struck.

With Burma suffering the brunt of the Cyclone Nargis disaster, the country and its government were in the international spotlight. In investigating disaster diplomacy, it is legitimate to ask whether or not that might force the country to open up. Could the referendum or Aung San Suu Kyi be affected? Should they be affected?

As these questions were being raised, the international aid operation began. Burma's government initially declined international aid. Then, on 8 May, one week after the disaster started, the first UN airplane with aid landed. The next day, amid accusations of aid being impounded, Burma said that it needed international aid, but not international aid workers. On 10 May, the constitutional referendum proceeded across most of the country, with the areas worst-hit by the cyclone being exempted. Two days later, the first American aid airplane landed in Burma.

Burma's government announced the referendum results on 15 May. With a voter turnout stated as being in excess of 99 per cent, 92.4 per cent had voted in favour of the government's plans for a new constitution. Meanwhile, the aid effort continued.

Three weeks after the disaster started, on 22–23 May, UN Secretary General Ban Ki-moon visited the affected areas and held talks with Burma's leadership. As part of that visit, Burma's leaders announced that all international aid workers would be permitted to enter Burma while aid would be accepted from civilian ships. Burma also publicly thanked foreign medical workers who had assisted in the disaster zone. On 25 May, an international conference in Rangoon attended by 52 countries pledged aid to Burma as Burma requested US$11 billion to rebuild. Burma's Prime Minister told the conference that international aid will be accepted with 'no strings attached'.

Two days later, Aung San Suu Kyi's house arrest was extended for another year. In less than a month, any disaster-diplomacy possibilities had evaporated.

Part of the disinterest in Burma's government changing might have resulted from intense international pressure on the regime to make disaster diplomacy work. Regional and international media were rife with speculation that the government would be forced to open the country due to the post-cyclone humanitarian imperative and that might spell the end of the government's oppression and isolation. Meanwhile, France's Minister of Foreign and European Affairs, Bernard Kouchner, wrote that the world has legal and ethical obligations to provide aid against the wishes of the Burmese government (Kouchner, 2008), creating an intense debate regarding sovereignty in times of disaster.

To blame disaster diplomacy's failure on the outside pressure misses the root cause of the government's intransigence. Burma's government was not willing to warn their people that they faced a catastrophic storm, nor was it willing to assist people suffering from the disaster in the most efficient manner. To the government, maintaining its grip on power was more important than disaster risk reduction or disaster response. That is the true failure of disaster diplomacy in this case. Neither disaster nor diplomacy were necessarily a high priority for the decision-makers – even to save tens of thousands of lives of their own citizens.

Ten days after Cyclone Nargis hit Burma, on 12 May 2008 an earthquake struck Sichuan in southwest China. The death toll topped 40,000, including devastation wrought by schools collapsing, burying thousands of pupils and staff. Like Burma, China is not democratic, is heavily criticised for human-rights abuses, and is relatively closed to the outside world, through policies such as internet control and restricting access to specific regions. What disaster-diplomacy chances existed for China dealing with the outside world following the earthquake disaster?

Within hours, Chinese leaders were visiting the devastation and facilitating international relief efforts. Few media restrictions were evident in the reporting. International teams were welcomed to assist in the search and rescue as well as to assist the survivors. One of the big stories was the anger from the earthquake-affected population regarding shoddy school construction, mainly as a result of corruption, that led to the loss of a generation of youth in a one-child-per-family society.

Meanwhile, Taiwan's government and citizens immediately offered relief donations and assistance. China accepted the offers with thanks. This friendly exchange was different from nine years earlier. On 21 September 1999, an earthquake hit Taiwan, killing over 2,000 people. Taipei and Beijing had a spat regarding the provision of relief supplies. The conflict over aid was fuelled by Beijing's attempts to control international relief and to create an international role for itself (see also Weizhun and Tianshu, 2005) alongside Taipei's open distrust of any support, in words and in aid, coming from Beijing.

Why did the Sichuan earthquake reveal disaster diplomacy differences from Taiwan's earthquake disaster and the Burma's cyclone disaster? One difference is the nine years between the two earthquakes. China has been increasingly open to the outside world, in part to help flex its economic muscle and to engage with the world on its own terms. The timing of the Sichuan earthquake, just three months before the Beijing Olympics, also made a difference. China was experiencing human-rights protests during the Olympic torch relay and was coming under increasing international scrutiny regarding its democracy and human-rights records. To yield a positive spin for the Olympics, a clear incentive existed to manage the disaster appropriately and to be seen to be open about the dealing with the disaster.

Is it too cynical to assume that the disaster diplomacy was a public-relations exercise? As the international attention shifted away from the earthquake and with the Olympic Games in August judged to have been successful, any potential disaster-diplomacy gains vanished. Media control over the disaster increased, protests were stifled, and those speaking out regarding the school collapses were arrested. This pattern parallels Burma. A relatively short time-frame reveals that decision-makers have interests other than dealing with the disaster and other than creating diplomatic dividends from the ruins.

The May 2008 disasters show that disaster diplomacy succumbs to priorities other than dealing properly with disasters and creating diplomacy.

4.14 Island evacuation due to sea-level rise

Creeping environmental changes, also termed creeping environmental problems or creeping environmental phenomena, are incremental changes in conditions which cumulate to create a major catastrophe or crisis, with the crisis apparent only after a threshold has been crossed (Glantz, 1994a, 1994b). Creeping environmental changes such as desertification and salinisation of water supplies significantly impact all spatial scales. They frequently cross sub-national and international borders. That makes them useful cases for disaster diplomacy.

Examples are droughts in southern Africa from 1991 to 1993 (Holloway, 2000; section 4.4) and from 2002 to 2003; the water diversion from and drying up of the Aral and Caspian Seas (Glantz, 1999); and the impact of precipitation changes on Fouta Djalon, the headwaters in Guinea from where the Niger, Senegal, and Gambia Rivers start.

One global creeping environmental change is the long-term disaster of contemporary climate change, a significant proportion of which results from human-caused greenhouse gas emissions and deforestation (IPCC, 2007). In the history of greenhouse gas emissions, most have come from larger, more affluent countries. Much deforestation occurs in less affluent countries, but primarily to serve commercial markets in the larger, more affluent countries (Butler and Laurance, 2008). Smaller countries and communities, including many sovereign and non-sovereign islands and island groups, have contributed little to greenhouse gas emissions or to forest destruction, even when rates are normalised by GDP or per capita (IEA, 2009; Roper, 2004).

Yet many islands, especially their coastal zones, are under severe threat from the creeping environmental change of sea-level rise, resulting from melting ice from glaciers and polar ice sheets along with the ocean water expanding as it warms up. As one example, residents of the Carteret Islands of Papua New Guinea have already left their homes and moved elsewhere, seemingly due to sea-level rise (Connell, 1997). Other island communities such as in Alaska are planning moves inland, as they face increasingly rapid coastal erosion, partly occurring because less ice on the ocean permits more severe storms to occur (GAO, 2003).

Countries that might need to move much or all of their population elsewhere due to sea-level rise include Tuvalu, the Maldives, and Kiribati. That entails negotiating with other countries over the consequences of the climate change and sea-level-rise-related disaster, namely where any migrants would go and who would pay for the resettlement. That makes this case study relevant for disaster diplomacy related to creeping environmental changes (Kelman, 2006b).

If an island community or island country decides that complete resettlement is an appropriate option, the first decision that needs to be made is the timing of the migration. Should the people move as soon as possible, before it is essential to move due to the environmental changes? That approach would enable migration to involve appropriate planning, which might take years, but which would include preparing the new communities in which the islanders would be resettled, which is likely to involve a new country.

Alternatively, would it be more straightforward to garner support for a rapid resettlement after a major disaster, such as a storm flooding much of a low-lying atoll? That approach risks loss of life along with loss of possessions including cultural heritage artefacts. As well, the trauma of sudden migration and resettlement might occur because little planning would have been completed, or psychologically accepted, beforehand.

This discussion regarding extreme events that could severely affect the island communities to be resettled highlights a problem with the option for longer-term planning. The problem is that a disaster could strike at any time, forcing more rapid evacuation than is being planned for. Combining both approaches could be considered. That could entail developing plans to leave as soon as possible after an extreme event is threatening or strikes. The latter option, again, leaves the population exposed to the risk of fatalities and the loss of invaluable possessions.

Once the timing for the resettlement has been decided, then the disaster-diplomacy elements of this case study become more evident. The islanders need to decide where to go in order to create their new community or country. The negotiations over location are also likely to involve negotiations over the level of sovereignty or autonomy which the new community – or country – shall have.

From the islanders' perspective, it should be their own decision. Their country is being ruined, so if they wish to continue being sovereign, then that is their right. Simultaneously, other sovereign states have a right to continue owning their current land. Few precedents exist in the modern era of a sovereign state being ordered by the international community to give up some land without, at minimum, mutual consent regarding a negotiation process. That occurs, for example, when arbitration panels demarcate borders after a war, as noted for the Eritrea–Ethiopia case study (section 4.8). Border settlement processes through the international-court system are also

Empirical evidence 57

effectively mutual consent regarding a negotiation process and acceptance of that outcome.

No intimation exists from the countries threatened by sea-level rise that they would voluntarily give up their sovereignty. In contrast, some Pacific island diplomats are concerned that international discourse on the topic might be trying to force them to give up their countries' sovereignty if migration due to sea-level rise becomes a reality (McNamara and Gibson, 2009). Retaining sovereignty, and in particular retaining the threatened people's right to choose their own sovereignty pathway, might be the aspect that generates the most inter-state conflict within this case study. It might even preclude any disaster diplomacy outcomes, demonstrating how other factors supersede the humanitarian imperative in disaster diplomacy case studies.

Resolving this challenge is not straightforward. Multiple elements come into play regarding the right to sovereignty and rights to specific sovereign elements in governing a country.

For instance, different countries have different views on human rights. Tonga still retains whipping and death by hanging as possible crime punishments (Laws of Tonga, 1988). That compares to Australia and New Zealand which oppose capital punishment. If Tongans are forced to re-create their country elsewhere due to sea-level rise, but wish to retain their criminal code while settling in Australia or New Zealand, whose views on capital punishment should prevail?

Conversely, Kiribati has outlawed the death penalty while California retains it. If I-Kiribati migrants settle in California due to sea-level rise, should they be exempt from the death penalty in California's laws? Visitors from a jurisdiction without capital punishment are not immune from capital punishment when in a jurisdiction that supports it. Should forced migrants be accorded different rights?

Further complications emerge in arguing that resettled migrants have the right to retain their culture, language, and identity, whether sovereign or not. The act of moving in itself will change culture. As one example, it would be preferable to resettle islanders on land that is similar to, but more secure than, their current location. Other islands are an obvious choice. Yet that is unlikely to be feasible considering that most low-lying areas, including islands, would suffer similar sea-level rise fates as the islands being evacuated.

Many potential island candidates or coastal candidates for re-creating island communities or countries are already protected as environmental, tourist, or scientific havens. Examples are Australia's Great Barrier Reef islands and New Zealand's Kermadec Islands. Other locations are uninhabited because they are uninhabitable for large populations due to their size or resource constraints.

The Spratly Islands in the South China Sea (section 8.1) are an example. They have limited arable land and limited freshwater, as shown by the need for external assistance to maintain garrisons on some of the islands. In fact, most garrisons hold only dozens of soldiers at a time. Malaysia has built a small tourist resort on one of the Spratly Islands.

The Spratly Islands introduce further disaster-diplomacy concerns in that six countries claim them or parts of them: Brunei, China, Taiwan, Vietnam, Malaysia, and the Philippines. The competing interests are due to resources of gas, oil, and fish. Additionally, the islands are a strategic navigational and surveillance location. None of the disputants is likely to give up their claim easily.

A scenario could emerge where one of the claimant countries volunteers to give up its Spratly claim to the islanders being resettled – in exchange for a resource agreement. China and Taiwan are already embroiled in diplomatic competition around the Pacific island states (Biddick, 1989). They provide aid in exchange for the Pacific island state's recognition of either one China (Beijing) or else Taiwan's sovereign legitimacy (Taipei).

Given the difficulty of finding land that is similar to the islanders' current homes, major cultural changes will need to be accepted as part of the resettlement. Some of the island migrants are likely to resettle on the mainland of other countries. For the Pacific region, Australia and New Zealand are frequently touted as being the most likely candidates to provide some land. Possibilities such as Fiji, Indonesia, Japan, the Philippines, the Solomon Islands, Vanuatu, and the USA should also be considered.

For the Maldives, India and Sri Lanka might be considered to be obvious candidates as well. Nearest proximity, though, should not be assumed to be the criterion likely to lead to the most resettlement success.

While major cultural changes need to expected, major parts of the islanders' governance and culture might have the potential for being retained despite resettlement. Parallel and complementary justice systems for indigenous people operate in Canada (for example, Andersen, 1999; Baskin, 2002) and New Zealand (for example, Gibbs and King, 2002; Goren, 2001). As noted earlier, conflicting views of human rights could cause a concern.

Parallel currency systems could also be maintained. That would emulate situations that involve bartering and local currencies alongside national (for example, American dollars) or supra-national (for example, euros) currencies. Local Exchange Trading Systems exist in Ithaca, New York (Crowther *et al.*, 2002) and Skye, Scotland (Pacione, 1997). Non-national currency forms tend to lead to further advantages for the community, especially those that are isolated from the mainstream of economic activity (Croall, 1997; Williams, 1996). Many border towns use two currencies, their national currency as well as the currency just across the border. That encourages customers from both sides of the border and to avoid the losses that are incurred when changing money in banks. Enniskillen in Northern Ireland near the Republic of Ireland's border has long functioned in that manner.

Consequently, methods exist for retaining as much self-governance and culture as possible, despite resettlement. But no islanders should be under the illusion that everything about their current life, lifestyle, and livelihoods could be retained, even with full sovereignty in the new location. Additionally, non-territorial aspects of statehood are feasible as long as both the migrants and the host state or states are amenable and can cooperate. This form of disaster diplomacy might even create new forms of statehood or pseudo-states, especially in the context of the increasing inadequacy of the nation-state witnessed over the past century (Bereciartu, 1994).

An alternative to using existing land for the resettled islanders would be creating new land. The Spratly island of Layang Layang, which is Malaysia's tourist resort, was artificially created by filling in the shallow sea between two reefs. The Palm Islands off Dubai's shores were built to create new residential, leisure, and entertainment areas. If island creation is done for profit, why not for a people and a country?

Islands could be built that imitate the islanders' current homes, but which are high and wide enough to prosper despite sea-level rise and associated environment changes, including extreme events. Some of the legalities of artificial island states are detailed by Kardol (1999). Even more challenging, and potentially more dangerous and more legally contentious, would be constructing large mobile islands that would drift the world's oceans as mobile sovereign states.

One legal aspect is the Montevideo Convention (1933) covering the rights and duties of states. Article 1 defines a state according to having four criteria: 'a) a permanent population; b) a defined territory; c) government; and d) capacity to enter into relations with the other states'. Mobile states indisputably possess a), c), and d). A drifting island could potentially be challenged as having the status of a state for not having b). A mobile territory could still be a 'defined territory' or, since the state's ocean territory is continually changing, it might be argued as being a 'changing territory'. Because sovereign states are usually not keen to open established international conventions for modifications, in case other proposals are made for changes at the same time, it would be for international courts to set a judgment for a specific case in order to create a precedent.

For constructing territory, many other questions remain. Who would pay for construction and maintenance? How would territorial disputes be resolved? Examples are territorial water ownership from constructed static islands and sovereignty over marine resources around mobile islands. Could an island culture and state be viable in a mobile setting or is human culture too dependent on fixed land? Would lessons from nomadic people (for example, Markovits et al., 2003; Salzman, 1982) be helpful in establishing mobile island states? For islanders not used to nomadism, significant cultural changes would result as a consequence of trying to create mobile island states.

Questions also remain regarding the abandoned islands. If an island country is entirely evacuated, but the islands are submerged only at the highest tides, who owns the fishing and mineral resource rights in the surrounding seas? Could those rights be sold? Do the answers to those questions change if a sovereign state is disbanded rather than re-created on new territory? How certain is it that islands will need to be evacuated, due to inundation (for example, Webb and Kench, 2010) or other climate-change-related reasons such as freshwater resources?

This topic of disaster diplomacy from the creeping environmental change of sea-level rise due to climate change is not unique to islands. Many coastal settlements could suffer similar displacement for similar reasons. Although non-island coastal settlements have an 'inland' to which they can move, some islands also have that option, especially larger hilly islands such as Puerto Rico and Fiji's largest island, Viti Levu. Yet that would still result in significant changes, both for the people who must move and for the people already living inland.

As well, people in countries such as Bangladesh might need to cross borders in order to find an inland on which to re-settle. That raises the spectre of so-called 'climate refugees' or 'climate change refugees' that has a large scientific and policy literature that, as Hartmann (2010) points out, often fails to address basic questions regarding the use of that terminology.

Disaster diplomacy therefore becomes invoked again with many of the questions from low-lying islands applying to other locales affected by sea-level rise. This case study raises challenging moral and legal questions, many of which are currently being tackled through a rapidly expanding field of investigation.

4.15 Disaster-casualty identification

One possibility for disaster diplomacy which has so far not yielded prominent outcomes is that of post-disaster foreign-casualty identification and treatment, followed by repatriation of the injured or the bodies. When citizens of one country are involved in a disaster in another country, international collaboration is required, even if the countries involved are not cooperating diplomatically. The same is true for any political entity, even if not sovereign states.

In many of the other case studies reported here, if citizens of one country were caught up in a disaster in another country, then that situation was not reported prominently and did not influence the disaster diplomacy. Examples could have been Americans in Bam, Iran during the 2003 earthquake (section 4.2); Cubans or Iranians in New Orleans during Hurricane Katrina (section 4.12); Indians or Pakistanis in the earthquake-affected areas of Pakistan or India respectively (section 4.9); and Greeks and Turks in the 1999 earthquake of the other country (section 4.7).

The absence of reporting of such instances does not mean the absence of such incidents. That does indicate the lack of influence on higher-level diplomatic activities.

The literature on disasters involving citizens from multiple countries indicates the inherent challenges in international disaster-casualty identification. Two principal examples are provided here, followed by vignettes of other case studies, based on the available scientific literature. The first main example is the 27 March 1977 plane collision on a runway in Tenerife in the Canary Islands,

Spain. The second main example is the 26 December 2004 tsunamis, which were explored in sections 4.10 and 4.11.

Neither case study reported prior enmity amongst the international authorities dealing with the situation. Both case studies are explicitly described as research on international cooperation for disaster-casualty identification when multiple countries' citizens are involved.

The Tenerife airplane collision involved one American jet and one Dutch jet, killing 583 people with 61 survivors, many of them badly injured. Thirteen of the fatalities survived the crash, but died afterwards. Dutch and American police and forensic teams soon arrived to deal with the bodies. All 248 people on board the Dutch plane were killed with van den Bos (1980) reporting that 'there was not one body that could be shown to relatives' (p. 266). His team was able to identify all bodies, but it took three months. Van den Bos (1980) also notes challenges in dealing with the Spanish laws in terms of burial, which were different from the Dutch system.

In this instance, no enmity is implied amongst the authorities involved. Cooperation meant that the time needed for dealing with the circumstances was available and families of the bereaved could be assisted as much as feasible. That is especially with regards to knowing that the body that they were burying was the correct body for their grieving. In circumstances with countries or authorities that are less collaborative, the potential exists for additional trauma for the bereaved in navigating the political conflict while coming to terms with their bereavement. Alternatively, the potential exists for collaboration to assist those suffering which might forge links between the enemies.

Solheim and van den Bos (1982) record the operational lessons from their various experiences in disaster-casualty identification. They describe how to ease protocols for collaboration amongst states for disaster-casualty identification. None of the cases that they describe involves enemy states. Irrespective, no provisions are suggested in case of concerns about sharing information across borders, difficulties in getting visas for teams from the nationalities affected, or authorities' disinclination to cooperate.

Regarding the 26 December 2004 tsunamis, Scanlon (2006) details international cooperation, and tensions, for body identification, mainly in Thailand but also including Sri Lanka and Indonesia. The challenges emerged, not due to enmity or diplomatic difficulties, but due to different cultural and legal protocols for dealing with bodies. Examples were disagreements amongst Europeans from different countries and international teams' desires to focus on nationals from their own countries.

In Thailand, further disagreements arose due to the initial decision by the Thais to separate Thai and non-Thai bodies. That, in effect, ended up separating Caucasian-appearing and Asian-appearing bodies. Singapore, Japan, and Australia objected to that separation, noting that many of their citizens are Asian but are not Thai.

Most disagreements were soon resolved through discussions leading to developing common protocols. Enforcing those protocols was not always straightforward. Scanlon (2006) notes how creative police could be in seeking specific but varied solutions for obtaining fingerprints for body identification. Of particular importance was agreeing to exhume mass burial sites in Sri Lanka in order to proceed with identification. As with the Tenerife airplane collision, no concerns are mentioned regarding diplomatic disagreements or nationals of countries with conflicts running into obstacles due to those conflicts. Instead, the disagreements were operational and were generally resolved through discussions and agreeing on common protocols.

One other incident relevant to this case study is the suicide bombings that hit two hotels hosting Israeli tourists around Taba, Egypt on 7 October 2004, as described by Karp et al. (2007). Due to the remoteness of the location from medical facilities in Egypt as well as the people who were targeted, the main medical response was from Israel, with the casualties being taken to Israeli medical facilities. As Karp et al. (2007: 107) note, 'Although there is a peace agreement between Egypt and Israel, Israeli military and civilian teams had to obtain an official special waiver from Egyptian authorities to cross the border, delaying their arrival at the scene of the two bombings'.

Karp et al. (2007) focus on the treatment of the casualties. No intimation is given that the response should have or did contribute to or inhibit collaboration between Israel and Egypt. They did not even mention the citizenship of the casualties. Instead, it was assumed that the nearest responders would be involved and it was simply a logistical, not diplomatic, challenge regarding visa-related protocols. That logistical challenge was overcome by granting the Israeli responders a special waiver to enter Egypt, so that the casualties could be treated and transported.

Many other prominent disasters could be further explored for this case study. The 11 September 2001 terrorist attacks in the north-eastern USA killed citizens from numerous countries in the World Trade Center and on board the hijacked aircraft. The *Estonia* ferry sank on 28 September 1994 in the Baltic Sea, killing 852 people, mainly from several Baltic countries. Bombings targeting tourists in Bali, Indonesia in 2002 resulted in casualties from five continents. Casualties from four continents resulted from another series of bombings targeting tourists in Bali in 2005.

For the 2002 Bali bombs, Scanlon (2006) notes that, after negotiations, Australian police led the body identification because so many of the casualties were Australian. Additionally, 'Korea, Japan, the Netherlands, Sweden, and New Zealand assisted with the dead. Britain and the United States assisted with the investigation. But everything was done under a two-country agreement between Indonesia and Australia' (Scanlon, 2006: p. 57). Scanlon (2006) also describes that the USA declined assistance for disaster-casualty identification after Hurricane Katrina (see also section 4.12) while Pakistan did so after the 2005 earthquake (see also section 4.9) 'but most victims were citizens of the country impacted' (p. 61).

Disaster-casualty identification has disaster-diplomacy potential. The case studies reported indicate a focus on dealing with the disaster and assisting those needing help, rather than letting international politics interfere with that or seeking positive political outcomes. That is fair to the casualties and emergency-response teams who, in most circumstances, are unlikely to be concerned about the diplomatic implications of their work, instead focusing on saving lives.

4.16 International vaccination programmes

Throughout human history, disease has tended to be one of the most lethal types of disaster, killing millions of people each year (WHO, 2003). Basaltic flood, volcanic eruptions and strikes from astronomical objects such as comets and meteorites have the potential to exceed, in one event, the toll from several years of disease data. Such a situation would provide a useful disaster-diplomacy case study to see if enemies could collaborate regarding a global disaster threat.

Kelman (2003) examined how monitoring and warning systems for astronomical objects threatening the Earth could, but do not seem to, lead to disaster diplomacy outcomes. Consequently, given the lethality and global prevalence of disease, it is useful to explore whether or not 'disease diplomacy' exists and its contributions to disaster diplomacy work. The focus for disease diplomacy here is on vaccination programmes, in terms of international disease-eradication efforts and providing vaccines to enemies.

Other forms of disease diplomacy and health diplomacy exist too. Formal international health diplomacy dates back to at least the mid-nineteenth century (Fidler, 2001). A contemporary example is the UN's World Health Organization's (WHO) 'Health as Bridge for Peace' programme. It is premised on health workers being able to create peace through providing health care in post-conflict situations (Galtung, 1997).

Transboundary disease surveillance and control seeks to identify and curtail pandemics, often requiring cooperation amongst states in conflict. For example, during the 2005 Avian flu outbreak, the USA, which under President Bush from 2001–9 did not have an overly strong or constructive relationship with the UN, proposed mechanisms for international cooperation through the UN to tackle the disease (Dobriansky, 2005). Meanwhile, Israel and Jordan held talks on responding to the disease if it reached the region, although Palestinians were not invited.

Disease-eradication programmes cover efforts to collaborate internationally in order to eliminate diseases. The ethics of eradicating diseases has been questioned (Caplan, 2009), but WHO runs disease-control programmes which require cooperation from all countries.

Smallpox was declared eradicated on 8 May 1980, with the programme benefitting from multilateral support during the Cold War, even if the funds requested for the programme were rarely available (Barrett, 2003; Henderson, 1999). Other diseases targeted for elimination include dracunculiasis (Guinea worm disease; Hopkins and Ruiz-Tiben, 1991) and poliomyelitis (polio; MMWR, 2006).

Not all such programmes are affiliated with the UN. Providing vaccines to enemies at a bilateral level has been seen for North Korea (section 4.5). Another prominent case study is advocating for the American government to use vaccine diplomacy more actively. Hotez and Thompson (2009) argue that addressing diseases around the world should be part of the USA's foreign policy because it will lead to overseas populations becoming more friendly towards the USA. They also present the economic benefits of programmes to control and eliminate disease, making their proposals cost-effective foreign aid.

While their discussion is not deeply rooted in development and poverty literature, it provides a useful baseline for considering disease-diplomacy questions. In particular, how much of an impact would medical diplomacy have on the USA's stature around the world if it happens in parallel with less collaborative measures? One example of the latter is the takeover of city centres by American multinational corporations, especially food companies offering unhealthy menus. Another example is the American government continuing to spend orders of magnitude more money on defence and weapons than on development including medical diplomacy, as Hotez and Thompson (2009) highlight.

It is also possible that such discussion is spurious. Could most people be grateful for the assistance that they see day-to-day without concerning themselves about wider globalisation, politicking, and expenditures?

The proposals for the USA's medical diplomacy are based on tangible results from the past regarding international vaccination programmes. Hotez (2001a) starts in the early nineteenth century when the English inventor of the smallpox vaccine, Edward Jenner, became such a hero that he mediated prisoner exchanges between England and France. Then, during the Cold War, the polio vaccine was developed and distributed internationally through American–Soviet collaboration (Hotez, 2001a, 2001b).

Since then, Hotez (2001a, 2001b) writes, ceasefires in conflicts have been negotiated to permit polio and other vaccination campaigns under the UN to proceed in several countries. Examples are Afghanistan, the Democratic Republic of Congo, Liberia, and Sierra Leone. In Sudan in 1995, a ceasefire of six months was negotiated to combat drancunculiasis, representing the longest ceasefire in the conflict up until that time.

None of the case studies led to conflict reduction or resolution. Hotez (2001a: 865) does not overplay the results: 'No evidence is currently available to suggest that aggressive vaccination together with its expected decrease in childhood mortality would pre-empt an armed conflict.' The parties involved were content to hold a ceasefire for the vaccinations and then to pick up arms again afterwards.

Arguably, the vaccine-related ceasefires were successful because no attempt was made to link them to wider conflict resolution or conflict reduction. The combatants, potentially, had little interest in peace. Consequently, they permitted the medical diplomacy because they knew that it would not affect the violent conflict. That is not a failure of the vaccination efforts. It does indicate the limited impact of disease diplomacy, with Hotez's (2001a: 865) caution that 'it would be of interest to examine the long-term impact of childhood vaccination on a nation's foreign policy'.

As well, none of the case studies intimated more insidious motivations, such as the vaccinations providing a healthier population to draw upon for soldiers. Yet the vaccine diplomacy analyses do not acknowledge the possibility that a peace based on disease eradication could crumble once the disease-related goal has been achieved. That would emulate the hostilities recurring after the vaccination-related ceasefires ended.

In fact, Hotez's (2010) vision for improved American foreign policy towards and relations with Islamic countries based on medical diplomacy does not recognise that the medical diplomacy might fail if improved relations are deemed undesirable by the parties involved. That does not immediately preclude the attempts. That does suggest that greater awareness of the realism surrounding diplomatic interactions might indicate that less optimism is warranted for assuming that medical and health interventions must be positive for diplomacy.

This brief review of this vast case study across multiple dimensions suggests that most disease diplomacy efforts tend to avoid strident political paths. Successes are seen in the programmes moving forward and having an impact on the ground, at least in terms of health. Few tangible outcomes emerge beyond health. The focus on the health outcomes, rather than diplomatic outcomes, might lead to the successes witnessed by narrowing goals to target a specific disease, rather than trying to be all-encompassing by bringing peace to a region. Conversely, the final health outcomes desired are not always achieved, such as eradicating targeted diseases.

One possibility is that the health outcomes sought are too ambitious. Smallpox and its vaccine had advantageous characteristics that made the disease susceptible to eradication. Henderson (1999) emphasises that the smallpox vaccine is heat-stable and confers immunity to the recipient with one dose. Such characteristics are not repeated fully across the other diseases targeted for eradication. Nonetheless, the examples of dracunculiasis and polio show encouraging signs. Furthermore, little doubt exists that the programmes are suffering from a lack of financial support more than from a lack of diplomacy or moral support.

Financial support often emerges from political support, leading to the argument that international vaccination programmes are mistaken in focusing on disease rather than diplomacy. Trying to increase politicking to support a programme is a gamble, because political winds can change direction, leading to funding cuts.

Early attempts to tackle the HIV/AIDS epidemic in the USA in the 1980s were stymied by the conservative establishment led by President Ronald Reagan (Reeves, 2001). They saw HIV/AIDS as being a disease of homosexuals, while they subscribed to a certain sect of Christianity that tarnished

66 *Empirical evidence*

homosexuality as being sinful. That world view left little scope for taking a longer-term view of HIV/AIDS, to understand the causes and, on that basis, to hinder its spread.

Irrespective of how much health professionals would have wished to treat HIV/AIDS as simply a disease to be tackled without politics, from the beginning in the USA, the disease was interpreted to be political which affected the work of the health professionals. In this instance, the disaster of HIV/AIDS left little scope for disease diplomacy because the politics of sexuality exacerbated the disaster.

Therefore, the argument that international vaccination programmes are mistaken in focusing on disease rather than diplomacy might be fair in some circumstances and unfair in others. Deeper investigation into the disease case studies within a disaster diplomacy framing, across all forms of disease diplomacy, would contribute to sorting out strategic approaches for pursuing important health outcomes with the funding and political support required.

The final component to examine is whether or not any links or networks have been developed through disease diplomacy that led to non-disease-related diplomatic connections and eventually to non-disease-related diplomacy outcomes. Even when pursuing health outcomes as independently of diplomatic outcomes as feasible, further diplomatic consequences might manifest. The literature's discussion on such possibilities is limited, but it appears as if they are not being sought. Nonetheless, they are sometimes challenging to avoid and could be indirect disease diplomacy, operating subtly over the long term.

4.17 Summing up the case studies

The case studies presented in this chapter display a similar pattern with respect to disaster diplomacy: opportunities exist for disaster-related activities to create diplomacy, but those opportunities are rarely fulfilled. Despite wide varieties regarding the characteristics of the disaster-related activities and of the diplomacy, no case study emerges as being unambiguous disaster diplomacy.

That applies to pre-disaster case studies, such as Glantz's (2000) work on the Cuba–USA case study (section 4.6) and the planning for island evacuation due to sea-level rise (section 4.14). That also applies to post-disaster work, such as for Hurricane Katrina (section 4.12) and North Korea (section 4.5). A wide variety of geographic locations are covered as well, along with various manners in which diplomacy is expressed, such as through governments, international groups, and a country's citizens.

The diversity of case studies is thus a strength of the analysis. When the same outcome is observed across a multiplicity of situations, firmer conclusions can be established regarding disaster diplomacy. The following chapters articulate and analyse these conclusions.

Table 4.3 Summary of disaster-diplomacy case studies and research questions

	1 Disaster-related activities influencing diplomacy?	2 New diplomacy?	3 Legitimate diplomacy?	4 Long-lasting diplomacy?	5 Long-standing concerns addressed?
Burma and Cyclone Nargis	Yes	Yes	Yes	No	No
Casualty identification	No	Not applicable			
China and the 2008 earthquake	Yes	No	Yes	No	No
Cuba–USA	Yes	Yes	No	No	No
Ethiopia–Eritrea	No	Not applicable			
Greece–Turkey	Yes	No	Yes	Yes	Yes
Hurricane Katrina	Yes	Yes	Yes	No	No
India–Pakistan, 2001	Yes	Yes	Yes	No	No
India–Pakistan, 2005	Yes	No	Yes	Yes	No
Iran–USA	Yes	Yes	Yes	No	No
North Korea	Yes	Yes	No	No	No
Philippines	Yes	Yes	No	No	No
Sea-level rise and island evacuation	Yes	Yes	No	No	No
Southern Africa 1991–93	Yes	No	Yes	Yes	Yes
Tsunami, Aceh	Yes	No	Yes	Yes	Yes
Tsunami, Sri Lanka	Yes	No	No	No	No
Tsunami, other case studies	Mixture	Mixture	Mixture	No	No
Vaccination programmes	No	Not applicable			

To start the process of comparing and analysing these case studies more formally, Table 4.3 answers the research questions in section 3.3 for the case studies in this chapter. Meanwhile, the disaster-diplomacy hypothesis, which as noted in section 3.2 was formed and modified over the years in conjunction with the case-study analysis, is confirmed.

Across a wide variety of case-study types, forms, and locations, disaster-related activities have the potential to catalyse existing diplomatic activities in the short term, but not to create new diplomatic activities over the long term.

This short-term catalysis is not always witnessed. At times, disaster-related activities exacerbate conflict.

The affirmation of the disaster-diplomacy hypothesis and the answering of the disaster-diplomacy research questions for specific case studies provides a foundation for the following chapters. They examine the similarities and differences amongst the case studies through possible approaches for categorising case studies, developing quantitative analyses, and formulating qualitative typologies.

The evidence and analysis assist in further testing the disaster-diplomacy hypothesis and giving further depth and breadth to answering the disaster-diplomacy research questions. That reveals patterns across the wide variety of case studies.

5 Analyses and typologies for disaster diplomacy

5.1 Quantitative analyses

Several attempts have been made to run quantitative analyses to seek statistical relationships amongst different forms of disasters and different forms of conflict. Three studies are discussed here, as epitomising key points in the discussion that emerges from this approach: Brancati (2007), Burke *et al.* (2009) contrasted with Buhaug (2010a), and Nel and Righarts (2008).

Brancati (2007) explored how intra-state conflicts are affected by earthquake disasters. Data from around the world from 1975 to 2002 were analysed to check links amongst:

- Independent variables of earthquake Richter magnitude, population density, and earthquakes in bordering countries. The choice of Richter magnitude is surprising since that is not the favoured metric for earthquake magnitude used by seismologists and the scale was not designed for the largest earthquakes.
- Dependent variables of intrastate conflict based on three different variables from various datasets.
- Control variables of national gross domestic product per capita, rule of law, and mountainous terrain.

The results are interpreted as suggesting that earthquakes increase the likelihood of conflict. The worst situations were higher magnitude earthquakes in countries that had the lowest gross domestic products and pre-existing conflicts, especially in areas of high population densities.

A misapprehension is evident regarding why conflict might be affected by earthquakes or earthquake impacts, especially in poorly distinguishing between earthquakes and earthquake impacts. According to Brancati's (2007) argument, the earthquake disaster should affect conflict, rather than just the event of an earthquake. To have an earthquake disaster, much more than shaking is needed. In fact, Brancati (2007: 736) admits this limitation in writing, 'All of the results presented in this section may underestimate the true effects of earthquakes on conflict, however, since their impact depends on other unmeasured factors, such as the depth with which earthquakes occur in the earth's interior as well as the rock structure and soil composition of the area in which they strike'.

No explanation is given regarding why using real metrics of earthquake disasters, rather than only Richter magnitude, would necessarily increase 'the true effects of earthquakes on conflict'. They could just as readily decrease those effects. It is also surprising that the real aspects of earthquake destruction are neglected: building types and construction modes, especially the governance system for enacting, monitoring, and enforcing building codes. Without using appropriate measures for the earthquake impacts, it is unclear whether or not the analysis would have meaning.

Finally on the side of the earthquake variables, it is puzzling why Brancati (2007) uses the 26 December 2004 tsunami in Sri Lanka as an example of an earthquake exacerbating conflict. Tsunami genesis is correlated with submarine earthquake magnitude. The form of fault failure, especially whether or not vertical displacement occurs, is particularly important too.

For instance, three days before the tsunami disaster, on 23 December 2004, a moment magnitude 8.1 earthquake with a depth of 10 kilometres struck south of New Zealand, near Macquarie Island, generating only small tsunamis (USGS NEIC, 2005). Three months later, on 28 March 2005, a moment magnitude 8.7 earthquake with a depth of 30 kilometres near the location of the 26 December 2004 event also caused only relatively small tsunamis (USGS NEIC, 2005).

As well, bathymetry, topography, and morphology around the coastline significantly influence tsunami characteristics, especially as far from the earthquake's epicentre as Sri Lanka was on 26 December 2004. Fortunately, 26 December 2004 was not included in Brancati's (2007) quantitative analysis because it was outside the considered timeframe. The discussion of that case study in the paper adds to the misinterpretations regarding the connection amongst earthquake hazard characteristics and earthquake impacts.

That is further evidenced through the assertion that 'earthquakes are more likely to stoke feelings of frustration arising from relative deprivation than disasters with slow onsets, such as droughts' (p. 716), a statement given without empirical evidence. The explanation given is that 'Earthquakes ... unlike some other disasters such as famine, are exogenous to conflict' (p. 716). It is true that earthquakes are not usually related to conflict, but section 5.2.7 discusses attempts to create earthquakes showing that, on rare occasions, earthquakes can be related to conflict.

The main concern with Brancati's (2007) statement is that earthquake disasters have root causes in vulnerability (see section 3.1) which are strongly influenced by conflict. Conflict can undermine governance attempts to create, monitor, and enforce building codes. Conflict can inhibit the population's ability to prepare for earthquakes through emergency-management training and stockpiling post-disaster supplies. A fundamental misunderstanding is present in Brancati (2007) of the difference between earthquakes and earthquake disasters.

Brancati's (2007) analysis raises further questions. Post-tsunami Sri Lanka, where the conflict was exacerbated by the disaster (section 4.10), is analysed, as an example outside the study's time-frame. Post-tsunami Aceh, where the disaster contributed to resolving the conflict (section 4.10), is consigned to a brief footnote. That is particularly ironic given that Aceh suffered from both the earthquake and the tsunami whereas Sri Lanka suffered from only the tsunami.

With respect to Aceh, Brancati (2007: 720) writes that 'disasters can also kill important political leaders without which warring factions may lack the necessary leadership to conduct conflicts'. This discussion does not consider that the death of some leaders can also provide space for more violent, more radical warmongers who previously might have been held in check by more sensible superiors.

The focus of Brancati (2007) is on increased violence due to increased resource scarcity. The arguments for this interpretation in the paper acknowledge the literature that supports and promotes that viewpoint. Detailed explanations are given why resource scarcity can increase conflict. Links are made with other factors such as wealth and prior conflict. In contrast, Brancati (2007) does not discuss the critiques of that literature (for example, Peluso and Watts, 2001) nor the debates amongst the various schools of thought ('Exchange', 2003: Kahl, 2002).

When a correlation is found, causation mechanisms are needed to ensure that the correlation is real, rather than being spurious or coincidental. Brancati (2007) provides a possible causation, resource scarcity, and bases it in the scientific literature. Brancati (2007) does not recognise that the causation has the potential of being spurious because other scientific research demonstrates that, in some circumstances, resource scarcity does not always lead to conflict and can sometimes create cooperation.

Focusing on the numbers obscured some of the deeper meanings behind those numbers.

The study from Burke et al. (2009) displays similar characteristics. Using data from 1981–2002 for sub-Saharan Africa, they compared:

- Independent variables of monthly average temperature and monthly total precipitation over ½°×½° grid cells, which were then averaged over each country.
- The dependent variable of civil war per country per year, represented as a binary variable. It is defined as 1 if an internal conflict resulted in over 1,000 battle deaths in a country in a year. If not, then it is assigned 0.

Control variables are national per capita income and national levels of democracy.

Burke et al. (2009) report that increased civil war in sub-Saharan Africa correlates with higher temperatures much more than any correlations with rainfall. They then use temperature projections from climate change models to project expected increases in armed conflict and battle deaths by 2030, in the absence of interventions. The explanation given for the observations is focused on temperature affecting agricultural yields and hence subsistence livelihoods as well as national economies.

Burke et al. (2009) make the same mistake as Brancati (2007) in assuming that resource scarcity always and inevitably leads to increased conflict. As with Brancati (2007), this critique of Burke et al. (2009) does not imply that resource scarcity never leads to increased conflict. Instead, a deeper understanding of the science would acknowledge the debate in the literature and the empirical evidence showing different forms of conflict and cooperation changing in different ways under different circumstances of resource scarcities and resource abundances (Salehyan, 2008).

The discussion on the factors affecting conflict risk in sub-Saharan Africa is also disappointing. Economic fluctuations are highlighted along with climate change impacts on food and sea level. A statement is made that 'economic and political variables are clearly endogenous to conflict ... We interpret our result as evidence of the strength of the temperature effect rather than as documentation of the precise future contribution of economic progress or democratization to conflict risk' (Burke et al., 2009, p. 20673). That hardly does justice to the rich literature on factors affecting conflict risk in sub-Saharan Africa, such as Cold War proxy wars (for example, Bender et al., 1985; Collier and Sambanis, 2005); superpower support for despots (Ihonvbere, 1996); ethnicity, religion, and culture (for example, Dallaire, 2003); and greed for power for power's or money's sake (for example, Collier, 2000; Shawcross, 2000).

Overall, despite some ideas given in the paper, the question is not fully addressed regarding why temperature would be expected to have or not have a palpable influence on conflict given all these other factors. That causation discussion must be robust, irrespective of the statistical correlation. As with Brancati (2007), Burke et al.'s (2009) correlation is provided but without convincing explanation regarding a definite causation.

In a response to the paper, Sutton et al. (2010) raise this concern, noting that the paper's title implies causation but stating that 'the evidence presented is not substantive enough to warrant such a conclusion' (Sutton et al., 2010: E102). In replying to this letter, Burke et al. (2010b) repeat the mantra that resource scarcity must lead to conflict without considering the other literature. Burke et al. (2010b) also repeat the variables that they considered with regards to sub-Saharan African conflict, without acknowledging the variables that they did not consider.

Later, Buhaug (2010a) uses a wider data set to show that evidence does not exist to link climate with civil war in sub-Saharan Africa. Buhaug (2010a) also explains the other factors that appear to link with conflict more than climate. Burke et al. (2010a) responded by reiterating their same points, without fully addressing Buhaug's (2010a) analysis, and also by using ad hominem attacks. Burke *et al.* (2010a) admit that including more recent data in the analyses supports Buhaug's (2010a) conclusions, not the conclusions of Burke *et al.* (2009). Buhaug's (2010b) response repeats points that Burke *et al.* (2010a) do not address and describes the literature that Burke *et al.* (2010a) miss, enforcing the need to focus on causations rather than assuming that any correlation found must be real.

The third paper considered here, (Nel and Righarts, 2008), investigates how 'Natural disasters increase the risk of violent civil conflict' (p. 166). Their time period is 1950–2000 using:

- Explanatory variables of hydro-meteorological disasters, geological disasters, other 'types of natural disasters', all natural disasters which is the sum of the three previous categories, and 'rapid-onset natural disasters' which is the 'all' category without famine and drought. The data are from the EM-DAT database, discussed below.
- Dependent variables based on the UCDP/PRIO Armed Conflict Dataset (Gleditsch *et al.*, 2002; Harbom and Wallensteen, 2009). The onset of domestic armed conflict in a country each year is sought, examining both the year in which a disaster occurred and the year after.
- Intervening and control variables covering several development-related characteristics, with examples being infant mortality rate, GDP per capita, and GDP growth.

The analyses confirm the authors' hypothesis that the disasters examined in the paper increase the risk of the forms of violent civil conflict examined. They provide further insight into how different types of disasters, according to their categories, affect different forms of civil conflicts.

Nel and Righarts (2008) provide detailed explanations into the correlations, but are superficial regarding possible causations. Most of the brief causation discussion is linked to the intervening and control variables, highlighting underlying development concerns as leading to conflict potential with disasters as a possible trigger. Examples given are inequality and environmental stress, with limited explanation.

Some of the assertions are puzzling. They state that 'The distinguishing feature of earthquakes and volcanic eruptions is the large degree of destruction and upheaval that they cause, upsetting not only the collective action resources of the power-holders, but also of their challengers' (p. 175). Jonkman's (2005) quantitative analysis of disaster deaths (also using EM-DAT data – discussed below) does not show that earthquakes and volcanic eruptions are more fatal or affect more people than other forms of disaster.

While noting the continual challenges with disaster-death statistics, few sources point to exceptionally large numbers of volcano-related fatalities or displaced people during the twentieth century. Tanguy *et al.* (1998) calculate 221,907 volcano-related deaths since 1783 as a minimum value. That compares to the single flood of the 12–13 November 1970 cyclone and storm surge in Bangladesh killing at minimum 224,000 people (Sommer and Mosely, 1972).

This comparison neither precludes nor denies the possibility of a cataclysmic volcanic eruption in the future killing hundreds of thousands or millions of people. It does preclude such events over the time-frame that Nel and Righarts (2008) claim that volcanoes have been especially destructive.

Aside from the limited discussion of causation to explain their correlations, Nel and Righarts (2008) are limited by the data that they use. The disaster data is the Emergency Events Database EM-DAT (www.emdat.be) from the WHO Collaborating Centre for Research on the Epidemiology of Disasters (CRED) in Belgium. This database is the best that exists. Those involved in developing and maintaining the database deserve kudos for the continuing time and effort that they put into it. Unfortunately, as both CRED and Nel and Righarts (2008) accept, the data have several significant limitations (for example, see Peduzzi *et al.*, 2005; Quarantelli, 2001; Sapir and Misson, 1992; Toya and Skidmore, 2007).

In particular, the categories used by Nel and Righarts (2008), based on EM-DAT, display numerous errors. It is unclear why tsunami is categorised as a 'Hydro-meteorological' disaster. While some tsunamis are caused by rainfall-induced landslides, the most destructive tsunamis historically have originated from geological phenomena (earthquakes, volcanoes, and geologically related landslides) or from large objects impacting water on the Earth's surface, such as a meteorite striking an ocean (Bryant, 2008). Wildfire is legitimately divided into different categories by Nel and Righarts (2008), but drought is not, despite several different types of drought (Glantz and Katz, 1977; NDMC, 2006). Cyclones, hurricanes, tropical storms, storms, and typhoons are categorised as 'Windstorms' even though most fatalities tend to occur due to flooding (Jonkman and Kelman, 2005).

The problems with the classification might or might not affect the correlations. New analyses would be needed to check for any changes.

Additionally, one of the largest limitations of such quantitative analyses is the definition of 'disaster' used by EM-DAT:

> For a disaster to be entered into the database at least one of the following criteria must be fulfilled:
>
> - Ten (10) or more people reported killed.
> - Hundred (100) or more people reported affected.
> - Declaration of a state of emergency.
> - Call for international assistance.
>
> (EM-DAT website: www.emdat.be)

For the EM-DAT database to be manageable and reasonably accurate, these criteria are sensible and powerful. No fault is implied with EM-DAT in choosing clear, transparent criteria and for implementing those criteria which make the database useable and maintainable. Furthermore, EM-DAT is open and transparent regarding how their data are collected, verified, and reported. The website provides solid and lucid guidelines on how the data should be used for analysis. EM-DAT cannot be blamed for problems in how the data are applied and analysed. Those using the data need to understand the limitations and alter their interpretations accordingly.

The fundamental challenge in applying EM-DAT's data for quantitative analysis is that the EM-DAT criteria specifically exclude the types of disasters that might be most prevalent and that might affect the most people. Disaster literature uses several approaches and bodies of evidence to articulate this point.

'Invisible disasters' (La Red *et al.*, 2002) refer to the notion that most disasters are local, being the expression of the everyday construction of vulnerabilities (for example, Hewitt, 1983; Lewis, 1999; Wisner *et al.*, 2004). These disasters tend to be unseen beyond the communities affected. A dozen people killed in a landslide in a remote mountainous village might not make any news and will not result in a declaration of a state of emergency or a call for international assistance. Thus, it might not appear in EM-DAT's database. If nine people were killed out of a village of 90 people, all of whom were affected, then it would not qualify for EM-DAT anyway, despite being devastating for the village.

Can this concern be proven to exist in reality and to influence the analyses? Not really – apart from taking the words of La Red *et al.* (2002) – because, by definition, such disasters are not being reported. Invisible disasters rarely receive the attention given to events that can be recorded in an international database, even though the accumulation of invisible disasters might equal the impact of these larger events. As Lewis (1984: 178) writes, 'the small disasters, which recur much more frequently than the large ones and affect similarly large numbers of people in total, escape attention and escape international action'.

No evidence for that statement is provided, but it is a common-sense truism. By definition, the smaller disasters receive less attention. Meanwhile, the qualitative literature referenced throughout this book presents plenty of field evidence to support Lewis's (1984) assertion.

Another concern with quantitative disaster data that cannot be reflected in databases is 'silent disasters' (IFRC, 2000; Zaman, 1991). Disasters with some of the worst impacts occur with little fanfare or publicity. Epidemics, especially those happening over a long time-frame such as HIV/AIDS or malaria, are particularly highlighted as silent disasters. They are highly relevant to disaster diplomacy (see section 4.16). Yet they consistently fail to attract the media spotlight, fail to trigger the flow of donor money into primary health care, and fail to be considered in databases of disaster 'events' because they are chronic, ongoing disasters.

In terms of annual mortality, disease has so far exceeded all other forms of disaster since several million children are killed every year (WHO, 2003). Yet EM-DAT's figures have fewer than 500,000 people killed by all disasters in every year for more than the past twenty years. Hoffman (2003) describes 'The hidden victims of disaster' (p. 67), referring to the most marginalised people being affected the worst by disasters, while receiving the least attention and the least assistance.

With invisible disasters, silent disasters, and hidden victims being so engrained in, and supported with plenty of evidence by, the qualitative disaster literature, most quantitative studies will experience severe limitations. That is not because the databases should be blamed, but because the databases by definition cannot collect these data.

That does not mean that the databases are pointless. Instead, they should be accepted as providing only a certain type of data, exactly as EM-DAT discusses. That type of data might not be appropriate for running correlations. To paraphrase Box and Draper (1987: 424) referring to quantitative modelling, essentially all disaster-related quantitative correlations are wrong, but some might be useful.

More appropriate interpretations of the available data and of the data limitations are found in some studies examining the relationship between climate change and armed conflict (for example, Buhaug, 2010a, 2010b; Buhaug et al., 2008, 2010; Nordås and Gleditsch, 2009). The evidence presented by those authors leads to their conclusion that, so far, little evidence exists that armed conflict is an important consequence of climate change.

In these analyses, they try to attribute potential causes to any changes in the correlations witnessed: 'intensification of natural disasters, increasing resource scarcity, and sea-level rise' (Buhaug et al., 2008, p. 2). These three processes are affiliated with five mechanisms:

- *Political instability*: Stable democracies are seen as the governance system that most avoids violence.
- *Poverty and economic instability*: While poverty and loss of livelihoods would not inevitably lead to armed conflict, the role of economic oppression in fermenting and perpetuating war is recognised.
- *Social fragmentation*: Again, various views are acknowledged. Different ethnicities are a potential spark in conflict, but views and evidence vary widely in the literature regarding the link between ethnic heterogeneity and violent conflict.
- *Migration*: The complexities and debates are explored of migration as both a cause and an effect of conflict, environmental degradation, and their links. Similarly to the previous two points, conflict potential frequently exists for migration that is linked to climate change (rightly or wrongly; see Hartmann, 2010). That result is not inevitable.
- *Inappropriate response and other contexts*: This discussion covers efforts from reducing greenhouse gas emissions to building dams to deal with more volatile water resource patterns. The overlaps with disasters and armed conflict are involved. The authors recognise the challenges of linking causes and effects along with non-scientific motivations in trying to

describe correlations and causations. For instance, they label as 'highly problematic' any suggestion of a 'near deterministic relation between the environment and armed conflict, thereby relieving the main actors of their own responsibility' (Buhaug et al., 2008: 29).

Their analyses are not confined to these elements or to solely elements that can be quantified and for which data exist. Wider contexts must be considered and input into the links. Examples that they consider are poor governance, inequalities, and population-related pressures covering high population densities, high population growth rates, and large percentages of youth. Their empirical review of studies involving population variables is clear that a large variation exists regarding the impacts of population-related variables on the links between climate change and armed conflict.

The discussion here of the limitations of the quantitative analyses has been conducted in a relatively qualitative manner. As such, the criticisms might be unfair by expecting too much from scoped studies that are constrained by the data available. The continual focus on causation rather than correlation might be doing a disservice to those authors whose scope does not include in-depth analyses of the causations, because calculating and interpreting the quantitative correlations require substantive effort, enough for a single paper.

That is, the correlations are potentially solid science even if drawing policy and practice conclusions from those correlations would be dangerous. The reason is that to use the correlations for action, adequate explanations of the causes are needed so that actions address the causes of the correlations observed.

This critique does not preclude quantitative studies nor does it suggest that quantitative studies could never be used for policy and practice. Instead, studies following good practices must be encouraged, especially when data and analysis limitations are admitted openly and are considered when interpreting the results. In particular, increased understanding of the error bars in the data and the results, along with sensitivity analyses to any errors, would be helpful in delineating the applicability of any calculated results. Continuing efforts to collect and refine data should also be encouraged.

Finally, mixed-methods approaches that seek to combine quantitative and qualitative analyses would be useful for cross-checking the results and for aiming to overcome the limitations of each method. Next, qualitative typologies for disaster diplomacy are examined.

5.2 Qualitative typologies

The case studies in Chapter 4 provide enough material to start to seek patterns amongst them, irrespective of the case studies' disparities. They also provide material to identify common as well as dissimilar characteristics amongst the case studies. That can help in moving towards answers for the 'how and why' questions surrounding disaster diplomacy (section 3.3).

The first such systematic attempt was done by Comfort (2000). She used the three case studies of Glantz (2000) looking at Cuba–USA, Holloway (2000) detailing the 1991–93 southern African drought, and Ker-Lindsay (2000) exploring Greece–Turkey following the 1999 earthquakes.

5.2.1 Complex adaptive systems

Comfort (2000: 280) described the conceptual model of 'complex adaptive systems', based on Holland (1995), to be applied to disaster diplomacy:

> This model focuses on the transition in different states of evolving social, economic, and political performance ... it recognizes that social systems engage, to varying degrees, in continuous learning and self-organization in reciprocal interactions with the environments in which they are embedded ... Literature addresses two basic issues regarding [complex adaptive systems]: 1) the conditions under which they emerge and function; and 2) the actual properties and mechanisms which characterize their operations. The two issues are interactive.

From Holland (1995) as discussed by Comfort (2000), complex adaptive systems have four properties and three mechanisms that need to be applied to disaster diplomacy and to the case studies.

The four properties of complex adaptive systems are non-linearity, diversity, flow, and aggregation.

Non-linearity is 'the condition in which small changes in a system's performance over time produce large differences in outcome, reflects the shift in energy and action within the component units of the system toward accomplishing a shared goal' (Comfort, 2000: 282). A political example comes from Canada in May 2005. A single politician decided to switch political parties, representing a small change. A few days later, on 19 May 2005, that prevented Canada's minority government from falling in a no-confidence vote, representing a large difference in outcome. A disaster example is one individual on a beach recognising the impending signs of a tsunami and giving a credible warning, representing a small change. That led one beach in Sri Lanka to be cleared just before the 26 December 2004 tsunami struck, with dozens of lives being saved, a large change (Cyranoski, 2005).

Diversity 'acknowledges that specific types of individuals or units may respond differently to the same events in the flow of ideas and actions, and interact accordingly to generate new flow among the components' (Comfort, 2000: 282). Diversity is represented in disaster diplomacy by the many parties involved, encompassing diplomats, politicians, international organisations, non-governmental organisations, the media, the private sector, and individuals.

Flow is the third property as 'the current of actions, materials, ideas, and people through a common arena that energises interaction among the individual units' (Comfort, 2000: 282). Material flow occurs when disaster-relief supplies are transported to a disaster-affected location or when diplomats transport security equipment for political negotiations. Information flow occurs with people tweeting (using twitter), emailing, phoning, texting, or sending post from a disaster-affected location or when diplomats transport documents for secret political negotiations in a conflict zone.

The final property is aggregation which 'represents the capacity for individual units to interact in a recurring pattern to accomplish a shared goal' (Comfort, 2000: 281). For example, the UN uses special appeals for donations following disasters in order to aggregate states' resources for disaster relief. Coordination and efficiency are other reasons for the appeals. Diplomats working through international organisations such as the UN could also represent aggregation in that the UN ostensibly shares a common goal of working towards a better, more peaceful world.

The three mechanisms of complex adaptive systems are tagging, the internal model, and building blocks.

Tagging 'facilitates the process of matching a unit seeking assistance with a unit providing assistance' (Comfort, 2000: 282). In some post-disaster scenarios, rescue units try to match their expertise with collapsed buildings that have not yet been cleared or that are known to have trapped people. In international diplomatic cooperation, states with mutual interests or with tradable commodities or services try to match up to seek agreements.

The internal model 'reflects the set of shared assumptions upon which reciprocal actions among components of the system are based' (Comfort, 2000: 283). For disaster-diplomacy situations, two governments might share ideas or assumptions about each other's amity or enmity towards the parties involved, including each other.

The final mechanism is building blocks which 'are the elemental units of performance that are used in creating a complex set of recurring interactions, such as communicative acts' (Comfort, 2000: 283). For disasters, meteorological stations are units that monitor raw weather data. They then process, interpret, and communicate those data, for warnings in the short term or for establishing trends over the long term.

Table 5.1 is from Kelman (2006a). It is constructed from the description of the properties and mechanisms, based on Comfort (2000), of the case studies in order to contribute towards explaining the disaster-diplomacy outcomes that are witnessed.

As a first attempt at a cross-case-study comparison, the results are helpful. The limitations of applying complex adaptive systems to this form of analysis are especially evident. Complex adaptive systems is a highly mechanistic approach. To a large degree, it treats systems of international relations as machine-like processes. That rarely does justice to the multi-faceted interactions and complexities witnessed amongst the mixture of parties that become involved in disaster-diplomacy cases (see also Kelman, 2010).

Table 5.1 The complex-adaptive-systems approach applied to three disaster-diplomacy case studies

		Cuba–USA (Glantz 2000)	Greece–Turkey (Ker-Lindsay 2000)	Southern Africa (Holloway 2000)
Properties of complex adaptive systems	Non-linearity	present	present	present
	Diversity	present	present	present
	Flow	limited presence	present	present
	Aggregation	present	present	present
Mechanisms of complex adaptive systems	Tagging	limited due to limited flow	recently present	present
	Internal model	limited due to limited flow	recently present	present
	Building blocks	present	recently present	present

Source: Kelman, 2006

For instance, personal amity and personal enmity influence disaster diplomacy. A strong success factor in the initial Greek–Turkish rapprochement, irrespective of the earthquakes, was the friendship between the two foreign ministers. Conversely, Fidel Castro's personality and history of conflict with the USA was a significant factor in Cuban–American antipathy when Castro headed Cuba.

Animosity and friendship can certainly be identified and described through complex adaptive systems' properties or mechanisms. Nevertheless, the properties and mechanisms do not explain from where and how the emotions arise. They also do not explain why specific case studies display specific attributes.

Identification is a needed starting point and that emerged directly from the original disaster-diplomacy question (section 3.3). With disaster diplomacy evolving towards the 'how and why' questions, the complex adaptive systems approach needs to describe how and why disaster diplomacy can fail or succeed due to personalities. Why are some personalities able to overcome enmity, irrespective of their personal feelings or external pressure? Why do others prefer to perpetuate animosity or to become trapped in a rut of conflict? As seen from the case studies, answers to these questions sometimes exist, but are not usually straightforward.

Another challenge of applying complex adaptive systems to disaster diplomacy is the properties. The properties can frequently be identified as being definitively present or absent in physical systems. They are inevitably present in international diplomacy at varying levels. Table 5.1 illustrates this challenge in that most cells are filled in with 'present' for both the properties and mechanisms.

The first property of non-linearity is always present to a high degree in politics and disasters. For disasters, neither hazard nor vulnerability match linearity. Hazards display non-linear pathways from hurricane tracks to earthquake frequencies. Vulnerabilities arise from, and are perpetuated by, complex political processes that affect diplomacy. People rarely behave straightforwardly, even where predictability is feasible, given the varying pressures of history, culture, politics, and media. Goals can vary from trying to make decisions that will enhance one's power base to following one's belief in a deity.

The second property of diversity is similarly present to a high degree all the time. Even if a specific tornado or volcano is involved in a specific disaster, the elements affected are diverse. Those elements include individuals, families, buildings, communities, livelihoods, jobs, social networks, and economies amongst others. Diplomacy by definition involves more than one player. It usually involves different sectors such as politicians, diplomats, media, and their overlaps.

For the third property, flow, even Kelman's (2006a) suggestion that flow is limited for the Cuba–USA case study is unduly pessimistic. Glantz (2000) is clear that significant flow is happening between Cuba and the USA at the scientific and technical levels. Remittances are another significant flow into Cuba and even into isolated North Korea (Kapur, 2005).

Cuba also has numerous international flights, especially tourist charters from several countries. Those flights include Americans who, when restricted from travelling to Cuba, circumvented their country's law.

For North Korea, even at the height of the country's isolation, a few tour operators ran tourist trips into the country. The North Korean government continued to run its own website and news agency to disseminate information to the outside world. North Korea maintained diplomatic ties with other states and is involved in international endeavours including the UN, in which North Korea is a member state. North Korea has signed various international treaties, such as the Convention on Biological Diversity – which North Korea signed two days before South Korea! Section 8.1 discusses North Korea and the Antarctic Treaty System.

All such activities lead to flow, even if the level of flow varies widely across case studies. Even physically isolated territories such as the UK Overseas Territories of Pitcairn Island and Tristan da Cunha implement multiple approaches for creating flow. Examples are short-wave radio, exchanges with passing ships, the internet, private ships visiting, and scientific expeditions.

The fourth property of aggregation is always present, as with the other properties. Also similarly to the other properties, the level varies. Following a disaster with international implications, some form of resource aggregation inevitably occurs at national and international levels. That could include joining and jointly directing funds or collaborating on the ground to avoid duplication in search and rescue operations. At the political level, government budgets pool resources for different sectors, including for diplomatic or international endeavours.

Much of this discussion and the examples seem to be self-evident, as further supported by Table 5.1. Again, that suggests that the properties of complex adaptive systems are not of significant use in comparing disaster-diplomacy case studies, even if the degree to which each property is present or absent could be described qualitatively or quantitatively. The more or less universal presence to a high level of all four properties in both disaster-related activities and in international affairs provides limited assistance in interpreting and developing explanatory and predictive models for disaster diplomacy.

Exploring the mechanisms for complex adaptive systems yields similar results. The presence or absence of the mechanisms tends to be more of an active choice decided by the parties involved. Tagging and the internal model, in particular, tend to follow rather than lead decisions. If those being party to disaster diplomacy wish to tag or to develop shared assumptions, then it is their choice to put the effort and resources into developing properties that would enable that.

As an example, these two mechanisms have at times been limited for Cuba–USA disaster diplomacy because both sides made active choices to impede them, especially with respect to tagging. Nonetheless, as noted earlier, the analysis in Table 5.1 by Kelman (2006a) is not entirely accurate because it does not fully account for the flow that does exist between Cuba and the USA. Once all Cuba–USA flows are considered, tagging and the internal model become more prominent for specific sectors or for specific instances.

Similarly, building blocks are present in all disaster-diplomacy case studies, but they are used only when active choices are made to use them. For the Greece–Turkey case study, the two governments initially decided to pursue rapprochement based on building blocks such as the Kosovo conflict, the EU, and shared interests in the Aegean. Following the 1999 earthquakes, Greeks and Turks, including the media, pushed disaster diplomacy forward using building blocks of shared experiences, similar cultural interests, and historical connections.

Conversely, for the Eritrea–Ethiopia case study (section 4.8), many building blocks existed. Examples are shared heritage, historical connections, the need to address a regional drought, and the humanitarian organisations' desire to address the emergency. The two governments chose not to use those building blocks to pursue disaster diplomacy. As with the properties of complex adaptive systems, the three mechanisms contribute to the discussion and understanding of disaster diplomacy. They do not fully explain or predict why or how disaster diplomacy does or does not manifest.

Consequently, less mechanistic approaches for trying to describe and compare disaster diplomacy case studies were developed by Kelman (2003, 2006a, 2007) and Warnaar (2005). The next sections describe, critique, update, further develop, and introduce new typologies and categories for other qualitative approaches for comparing the disaster-diplomacy case studies.

5.2.2 Qualitative typology: propinquity

One qualitative typology for disaster-diplomacy case studies is propinquity. That refers to neighbourliness or proximity of the disaster-diplomacy parties to each other, generally meaning the borders of geographic entities. Four main categories exist.

First, case studies where a single geographic entity is involved, so geographic overlap amongst the conflicting parties exists. Examples are post-tsunami Aceh and Sri Lanka (Le Billon and Waizenegger, 2007; Rajagopalan, 2006), famine in the midst of Sudan's civil war (Autesserre, 2002), and disasters affecting internal conflicts in the Philippines (Gaillard *et al.*, 2009).

The second propinquity category is parties that are physically connected. That frequently refers to states sharing a land border, such as India–Pakistan and Greece–Turkey.

Third, parties that are not physically connected, but which are near each other. For political and geographic entities, that normally refers to separation by a short expanse of water, but not sharing a land border. Japan and North Korea are in this category, as are Cuba and the USA.

The fourth propinquity category is parties that are not physically near each other. Examples are the USA and its disaster-diplomacy case studies with India, Iran, and North Korea.

Several limitations of the propinquity typology are immediately evident. Rivers and lakes frequently form international boundaries. They are midway between sharing a land border and not being physically connected. That situation is particularly of interest when borders are relatively long and a conflict zone is only one part of a large country. The Democratic Republic of Congo (DRC) provides two examples.

DRC's capital Kinshasa sits directly across the wide Congo River from Brazzaville, the capital of the Republic of Congo (ROC). Both capital cities have experienced violent conflicts over the years as regimes were toppled by or fought back against rebels. A hypothetical disaster-diplomacy case would exist regarding floods or flood-warning systems in this particular area of the DRC–ROC border encompassing the two capital cities.

The second DRC example is Goma, a town on the Rwanda–DRC border and on Lake Kivu which forms part of the Rwanda–DRC border. Following the 1994 Rwandan genocide, Goma hosted large temporary settlements of Rwandan refugees, complicated by many of the refugees being the genocide's perpetrators. Further complications included continuing internal violent conflicts in DRC (which at that time was called Zaire). Some of those conflicts originated from Rwanda and Uganda, with Goma as one invasion staging point.

On 17 January 2002, a volcano near Goma erupted sending lava into some of the refugees' settlements. The world responded swiftly with humanitarian aid, raising hopes for disaster diplomacy in the region, especially when some of the parties in the conflicts around Goma made concessions to assist with humanitarian relief. In the end, nothing materialised regarding the volcanic eruption influencing conflict reduction in that area.

These two DRC examples show the sub-national aspects of international borders. A country's international borders are not all the same, even with the same neighbour. Consequently, a disaster-diplomacy case study might have different neighbourliness characteristics along the same border.

Similarly, sharing an international border is not necessarily a sign of neighbourliness. North Korea and Russia share a short length of border, but disaster diplomacy between the two could occur at an international rather than local level. That means that Pyongyang and Moscow would be dealing with each other. Although the parties, being the countries, are physically connected, Pyongyang–Moscow interactions would be with parties that are not physically connected, instead being thousands of kilometres apart.

Another concern with the neighbourliness categorisation is the qualitativeness. Why should a 'short expanse of water' be any more dividing or less dividing than a short expanse of isolated land such as many desert or mountain range along an international border? Many Cubans died and many Cubans survived in trying to cross the short expanse of water to the USA. That gap has similar policing and survival difficulties as the desert and river along parts of the USA–Mexico land border which also experiences an influx from south to north. The difference between water and land is not as clear-cut as articulated in the propinquity typology.

The final main limitation with the neighbourliness characteristic is that entities involved in disaster diplomacy might be close geographically yet far apart culturally. Haiti and the Dominican Republic are both Small Island Developing States and share the island of Hispaniola in the Caribbean. Their histories and cultures are vastly different, especially given Haiti's French-speaking dominance compared to the Dominican Republic's Spanish-speaking dominance.

In February–March 2007, major floods struck parts of Bolivia. Rather than bringing different sectors in the country together, the disaster and the government's response to the disaster highlighted the disparity between élites and indigenous people across Bolivia. That occurred despite, or because, Bolivia's President at the time was indigenous. Even intra-state disaster diplomacy case studies can reveal and harden significant cultural and political differences irrespective of geographic proximity.

Conversely, in Colombia, communities within the Department of Meta have long had severe political conflicts over multiple topics. Floods are frequently devastating to many communities across the Department. One pioneering programme brought several communities in conflict together in order to deal with flood-risk reduction. It focused on protecting livelihoods, but also recognised that the process could build trust in other areas for further reconciliation (UNDP, 2004). The propinquity relationship is clouded. The communities are within the same geographic entity of Colombia but are clearly delineated, sometimes bordering and sometimes not.

Moreover, several propinquity categories could be present simultaneously. For North Korea and disaster diplomacy, South Korea shares a land border, Japan is nearby without a land border, and the USA is far away. If any internal dissent manifested, that would cover all four propinquity categories within a single case study.

The intra-state examples such as Aceh, Sri Lanka, Sudan, and the Philippines provide overlapping categories too. In some case cases, at the time of the potential disaster diplomacy, adversaries controlled specific territory with reasonably clear delineations. Tamil separatists were the de facto government in parts of northern and eastern Sri Lanka while the government of Sri Lanka governed the rest of the country. In Sudan, a north–south split in the country was a significant part of the challenge in bringing in disaster relief supplies.

Although intra-state conflicts are suggested as being the first propinquity category of a single political entity, Sri Lanka and Sudan could have been interpreted as the second propinquity category, at the time of their disaster diplomacy. That is, the two parties are physically connected with a de facto but unrecognised land border that is internal to a sovereign state.

For the Philippines, the Communist and Islamist guerrilla groups could be said to be in conflict with the USA as well as with the government in Manila. That is a different interpretation from fighting against the Philippines as a nation or as a sovereign state, or from fighting against the American presence and support for the Filipino government. Different levels of the propinquity categories appear when considering the disaster diplomacy roles of the guerrilla groups.

Table 5.2 provides examples of propinquity characteristics that are unambiguously evident in disaster-diplomacy case studies. Other forms of propinquity are present in many of the case studies.

5.2.3 Qualitative typology: aid relationship

Aid relationship amongst the parties involved in the disaster-diplomacy efforts is another disaster-diplomacy typology (after Warnaar, 2005). This typology has three main categories.

The first category is 'mutual aid'. That indicates that those involved in disaster diplomacy are helping each other on relatively equal terms. It could be because they are facing a common threat or because they have been affected by similar types of disasters. It could also be because they were affected by the same disaster and are cooperating with each other on that basis. Cuba and the USA's Gulf and Atlantic coasts face similar weather- and climate-related challenges, most notably hurricanes (Glantz, 2000). The same storm often hits both countries, with the 2005 hurricane season providing examples of Hurricane Dennis and Hurricane Wilma.

The second category of aid relationship is 'combined aid' which has two possibilities: first, parties in conflict coordinating aid to another party; second, several friendly parties coordinating aid to another party with which they have conflict.

Table 5.2 Propinquity evident in disaster diplomacy case studies

	One example of propinquity evident in disaster-diplomacy case studies
Burma and Cyclone Nargis	Not physically near each other
Casualty identification	Not physically near each other
China and the 2008 earthquake	Not physically near each other
Cuba–USA	Physically near each other
Ethiopia–Eritrea	Physically connected
Greece–Turkey	Physically connected
Hurricane Katrina	Not physically near each other
India–Pakistan	Physically connected
Iran–USA	Not physically near each other
North Korea	Physically near each other (Japan)
Philippines	Single geographic entity
Sea-level rise and island evacuation	Physically near each other
Southern Africa 1991–93	Physically connected
Tsunami, Aceh	Single geographic entity
Tsunami, Sri Lanka	Single geographic entity
Tsunami, other case studies	Single geographic entity (Maldives)
Vaccination programmes	Not physically near each other

An example of the first form of combined aid emerged from North Korea's droughts and floods after 1995. China, Japan, South Korea, and the USA collaborated to coordinate an aid package linked to other North Korean political topics. That occurred despite different levels and types of conflicts amongst these states as well as different levels and types of conflicts with North Korea.

An example of the second form of combined aid was different EU countries, the EU itself, and NATO providing a combination of bilateral aid and multilateral aid to the USA following Hurricane Katrina. That occurred despite the lingering enmity towards the USA and the UK amongst many EU and NATO states that actively opposed the 2003 Iraq War.

Donor-recipient is the third category of aid relationship. In these circumstances, one party involved is a donor, in that it assists another party in need. The party in need is the recipient of that aid; that is, it is being assisted. The donor–recipient role is not fixed, but can fluctuate depending on the circumstances.

Examples of donor–recipient disaster diplomacy are the USA offering aid to Iran following the 26 December 2003 earthquake and offering aid to Cuba following Hurricane Michelle in November 2001. The roles switched when both Iran and Cuba offered aid to the USA following Hurricane Katrina in August–September 2005. In the North Korea case study, North Korea has not yet provided aid to others.

Sometimes the role of donor or recipient is not inadvertent, but is deliberately defined. Following the 26 December 2004 tsunamis, India tried to maintain a role as a donor to other affected countries, despite India suffering over 10,000 dead. India soon accepted aid, deliberately deciding that it would be a recipient.

Limitations and overlaps are evident for aid relationship. In 'combined aid' situations, donor–recipient relationships exist too. For Greece and Turkey following the 1999 earthquakes, the case study began as donor–recipient after the earthquake in Turkey in August. Greece was the donor and Turkey was the recipient. The case study transformed into being mutual aid following the earthquake in Greece in September, when Turkey became the donor with Greece as the recipient.

Additionally, the categorisation of aid relationship might depend on how a case study is defined. Consider Cuba–USA and Iran–USA. If the case study is event based, then most interactions are donor recipient:

- Hurricane Michelle had the USA as the donor and Cuba as the recipient.
- The Bam earthquake had the USA as the donor and Iran as the recipient.
- Hurricane Katrina had Cuba and Iran as donors with the USA as the recipient, although the offered aid was never formally accepted.

Conversely, if the case study is based on geography, then both Cuba–USA and Iran–USA become a mutual aid case study, since they aimed to help each other at various times. This discussion reveals the challenges of categorising aid relationship when aid is not accepted. Case studies with potential donor–recipient roles or with offers of a donor–recipient relationship, such as Hurricane Katrina, could be contrasted with case studies where a donor–recipient relationship actually existed, such as Hurricane Michelle and the Bam earthquake.

Table 5.3 provides examples of aid relationships that are unambiguously evident in disaster-diplomacy case studies. Other forms of aid relationship are present in many of the case studies.

Table 5.3 Aid relationships evident in disaster-diplomacy case studies

	One example of an aid relationship evident in disaster-diplomacy case studies
Burma and Cyclone Nargis	Donor-recipient
Casualty identification	Combined aid (ii)
China and the 2008 earthquake	Donor-recipient
Cuba–USA	Donor-recipient
Ethiopia–Eritrea	Mutual aid
Greece–Turkey	Mutual aid
Hurricane Katrina	Combined aid (ii)
India–Pakistan	Donor-recipient
Iran–USA	Donor-recipient
North Korea	Combined aid (i)
Philippines	Mutual aid
Sea-level rise and island evacuation	Combined aid (ii)
Southern Africa 1991–93	Combined aid (i)
Tsunami, Aceh	Donor-recipient
Tsunami, Sri Lanka	Donor-recipient
Tsunami, other case studies	Donor-recipient
Vaccination programmes	Combined aid (ii)

5.2.4 Qualitative typology: diplomacy track

The third typology of disaster-diplomacy case studies is the diplomacy track witnessed for the disaster diplomacy. Kelman (2007) suggested that disaster diplomacy is conducted at three tracks or 'levels'.

First, government-led disaster diplomacy. The India–Pakistan case study after the 2001 earthquake is an example, because the states' governments were at the forefront of disaster-diplomacy efforts.

Second, organisation-led disaster diplomacy, involving groups that are not governments such as the UN, NGOs, the media, the private sector, lobby groups, and research institutes. Some disaster diplomacy of this form was evident following the Indian Ocean tsunamis of 26 December 2004 with the multiple efforts to promote peace in Aceh. The Eritrea–Ethiopia case study was also based on organisation-led disaster diplomacy when humanitarian agencies tried to open a corridor through Eritrea for aid delivery.

Third, Kelman (2007) described people-led disaster diplomacy in which grassroots support directs the efforts. That is often supported or given momentum by the media, as in the Greece–Turkey case study. Governments could also potentially push along people-led disaster diplomacy.

In developing this three-level typology of diplomacy 'levels', Kelman (2007) did not factor in the large literature covering different models of diplomacy tracks. Davidson and Montville (1981) defined two tracks for diplomatic interactions. Track-one diplomacy is for official routes, involving formal interactions with politicians, diplomats, and government officials. Track-two diplomacy covers 'unofficial, non-structured interaction' (p. 155) with examples being scientific interactions, cultural exchanges, and individual non-political visits. Davidson and Montville (1981) emphasise the importance of considering the two tracks to be complementary, mutually supportive, and needed for successful diplomacy: 'Both tracks are necessary for psychological reasons and both need each other' (p. 155).

Diamond and McDonald (1993) expanded these two tracks into nine tracks, coining the process of 'multi-track diplomacy'. They identified nine routes by which diplomacy occurs: 1) government; 2) professional conflict resolution; 3) business; 4) Private citizens; 5) research, training and education; 6) activism; 7) religious; 8) funding; and 9) public opinion/communication.

A major critique of this model is that it divides entities (for example, business) and purposes (for example, activism) as separate categories. That means that NGOs are divided according to their purpose(s), such as 'research, training and education' rather than 'activism'. Categorising diplomacy that occurs from, to, or through non-governmental armed groups would be challenging if that group's main activities are not specifically covered by one of the categories. Similarly, international organisations (for example, the UN or IFRC) and the media are not explicitly listed. They are implicitly present in several categories. Diamond and McDonald (1993) address some aspects of these

Table 5.4 Diplomacy tracks followed in disaster-diplomacy case studies

	One example of a prominent diplomacy track that was followed
Burma and Cyclone Nargis	Activism
Casualty identification	Private citizens
China and the 2008 earthquake	Government
Cuba–USA	Research, training and education
Ethiopia–Eritrea	Activism
Greece–Turkey	Private citizens
Hurricane Katrina	Business
India–Pakistan	Government
Iran–USA	Government
North Korea	Government
Philippines	Activism
Sea-level rise and island evacuation	Public opinion/communication
Southern Africa 1991–93	Government
Tsunami, Aceh	Professional conflict resolution
Tsunami, Sri Lanka	Government
Tsunami, other case studies	Religious
Vaccination programmes	Funding

critiques and their model of multi-track diplomacy has proved robust for many international affairs analyses.

Different combinations of these tracks can occur. Governments can deal bilaterally or multilaterally with each other or can be brought together by organisations. Organisations might deal directly with governments or with grassroots groups. People from one state can directly approach the government of another state. People can approach national and international organisations.

Table 5.4 provides examples of prominent diplomacy tracks in disaster-diplomacy case studies. Other tracks are present in many of the case studies.

5.2.5 *Qualitative typology: disaster diplomacy's purpose*

Multiple interacting purposes are often evident in disaster-diplomacy pursuits and outcomes, whether those are supporting or hindering disaster diplomacy. Trying to extract and understand disaster-diplomacy purposes is contentious, with views and analyses sometimes being expressed to conform to previously established partisan opinions.

The Cuba–USA example based on Fidel Castro's personality provides an example. A strong anti-Castro faction in the USA, particularly Florida, continually contended that Castro's policies and actions were taken only to perpetuate his own illegitimate and totalitarian dictatorship. This view suggests that the long history of Cuba–USA disaster diplomacy, and its long history of failure, emerge because Castro seeks to exploit disaster-related activities to prop up his power, irrespective of the ensuing harm to the Cubans and to the country.

Conversely, Castro's admirers tended to interpret disaster-related rapport between Havana and Washington, DC as evidence of Castro's willingness to put the political differences and the conflicts aside for dealing with disasters. That covered sharing information on hurricanes that were forming, accepting disaster-related aid for Cuba, or providing disaster-related aid to the USA. This view suggests that the common good of disaster-related activities frequently overcame Cuba–USA animosity due to Castro's willingness to work with the American government and vice versa. That was the case even where domestic political shenanigans in Washington, DC and Florida made it difficult for Cuba–USA disaster diplomacy to be fulfilled.

More cynical (or realistic) observers sought a middle ground, criticising both sides. They noted that American offers of disaster aid to Cuba conveniently highlighted Cuba's vulnerabilities and dependencies. The implied that Castro's leadership had not been kind to Cuba. Simultaneously, Castro would use similar reasoning for being willing to assist the USA.

Decoupling and linking the various reasons for disaster diplomacy from the complex range of factors, influences, and trade-offs accompanying both disaster-related activities and diplomacy is challenging. Several intertwined purposes are usually present at different degrees (see also Kelman, 2006a, 2010). An illustration of these complexities and of potential categories of disaster-diplomacy purposes emerges from Warnaar's (2005) and Kelman's (2007) discussions of Iran–USA interactions following the 26 December 2003 Bam earthquake. Five possible disaster-diplomacy purposes are summarised here, some material of which is from section 4.2, keeping in mind that many other potential purposes exist.

The first disaster-diplomacy purpose is survival of oneself. Iran recognised that foreign aid, including from the USA, would be needed following the earthquake. In this instance, 'oneself' refers to the political leaders in addition to the state. Parliamentary elections were due in February 2004. A poor response to the disaster could have hindered the government's re-election chances. Turning down aid could be criticised as being narrow minded and petty while Iranians were suffering. Yet Iran immediately refused aid from Israel, showing that this disaster-diplomacy purpose has limits.

The second purpose is that the disaster diplomacy would be of mutual benefit. In the months prior to the earthquake, Tehran and Washington, DC had been slowly reconciling some of their differences at a low level, as described by Armitage (2003). For the USA not to offer aid or for Iran not to accept that aid, whether or not that aid were needed, could have harmed the delicate détente which both sides were aiming to foster.

The third possible disaster-diplomacy purpose is long-term, global gains even if that has the potential for requiring short-term self-sacrifice. The earthquake response occurred in the wider context of regional and international relations. Iran was wrangling with the UN regarding inspections of nuclear-energy facilities and did not wish to appear closed to an international presence. The Bush presidency, focused on his re-election in the November 2004 American elections, and aware of slow progress in Iraq's and Afghanistan's post-war reconstruction, was keen not to have another regional destabilising influence, which could be presented by a major catastrophe in Iran.

Fourth, re-affirmation of old prejudices and enmity can affect disaster diplomacy interests. Iran stated that aid from Israel would not be accepted. This statement suggests that avoiding disaster diplomacy has as complex purposes as promoting disaster diplomacy. Iran's decision could have been made for internal political gain, due to entrenched animosity of the Iranian leadership against Israel, as part of nuclear-related regional politics (see Roger, 2005), or a combination along with other purposes.

The fifth possible disaster-diplomacy purpose articulated here is to prove humanitarianism. Neither Iran nor the USA was seen as being a compassionate state, regionally or internationally and both states suffer criticism for their human-rights record (AI, 2005; HRW, 2005). Acceptance of bilateral humanitarian assistance between unfriendly states provides each party with the opportunity to claim support for global humanitarian endeavours. That presupposes that the governments care about they are viewed with respect to humanitarianism.

Proving or disproving any of the above reasons for a case study is further complicated by factors beyond disaster diplomacy influences. Again, as illustrated by the Iran–USA case, five examples are given, drawn from Warnaar (2005) and Kelman (2007) with some reiteration from section 4.2.

Lack of political forethought is the first example. In early January 2004, apparently without consulting Tehran, Washington, DC offered to send a high-level political delegation as part of the relief effort. Iran declined. This response was perceived as being impolite, but was consistent with the American and Iranian positions with respect to each other before and immediately after the earthquake. The USA had tried to proceed too quickly with the rapprochement, creating too many links between the relief operation and the states' relations. Iran was not willing to accede.

The second factor is a country's internal ability to respond to a disaster. On 22 February 2005, an earthquake hit southern Iran killing more than 600 people. The USA offered aid which was politely declined. Iran's ambassador to the UN, Javad Zarif, noted that 'Iran did not refuse the help but said we can handle it domestically'. Aid from Algeria, Australia, China, Japan, the United Arab Emirates, and several international organisations was accepted, refuting that statement. Nelson (2010a) analysed a dataset of instances from 1982 to 2006 where disaster aid was refused.

The third element could be fear of disaster diplomacy as a distraction from the disaster. International politics and attempts to support long-term rapprochement could detract from immediate post-disaster needs.

Fear of disaster diplomacy because reconciliation is not desired is the fourth aspect. Both the Iranian and American governments perceived that they could gain politically – internally and internationally– by fostering some level of intergovernmental conflict. If disaster diplomacy occurred, political advantages could be lost from Iran's labelling of the USA as 'The Great Satan' and the USA's labelling of Iran as part of the 'Axis of Evil'.

Finally, the humanitarian imperative can be deemed to be important irrespective of its positive or negative impacts on diplomacy. Many organisations seek to separate disaster-related and diplomatic activities as epitomised by the Seven Fundamental Principles bonding the National Red Cross and Red Crescent Societies, The International Committee of the Red Cross, and the International Federation of the Red Cross and Red Crescent Societies (see also section 11.1). These principles include impartiality, independence, and neutrality.

Although governments rarely act entirely free from political drivers, a state's population can push governments into providing or accepting humanitarian assistance. That occurred to some extent in both Iran and the USA after the Bam earthquake.

5.2.6 Qualitative typology: type of state involvement

Interactions, relations, or connections amongst political jurisdictions are often assumed to be worthy of two distinct categorisations: inter-state and intra-state. Consequently, it seems prudent to examine disaster-diplomacy case studies by separating out inter-state and intra-state interactions and layering that on state-related and non-state related tracks of diplomacy (Table 5.5).

Examining case studies indicates that definitive categories are not always feasible. Considering the Greece–Turkey case study, Ker-Lindsay (2000, 2007) focused on state involvement and inter-state interaction, yet a key component of his analysis was the role of non-state parties such as media and the grassroots. That was especially the case regarding the influence of the media and public opinion on the state-level diplomats and politicians.

Mavrogenis (2009) also melded state and non-state involvement by examining the underdog culture in Greece plus Europeanisation as a driver towards peace between the two countries. Intra-state interaction, in terms of Greeks with political power working internally to change attitudes, was an important contributor to the change in attitude within Greek political circles. Similarly,

Table 5.5 Disaster-diplomacy examples for type of state involvement

	Inter-state interaction	*Intra-state interaction*
State involvement	North Korea	Philippines
Non-state involvement	Casualty identification	UK overseas territories working with each other and their NGOs for disaster-related activities; that is, island disaster para-diplomacy (section 8.2)

Ganapti *et al.* (2010) highlight the sector of disaster-related cooperation between Greece and Turkey, irrespective of the state-related connections of the organisations involved.

The 26 December 2004 tsunamis, covering all countries affected, demonstrate the diversity of parties and interactions occurring at various state and non-state levels. Each location with a tsunami-diplomacy case study yields ambiguities on its own. The conflicts in both Aceh and Sri Lanka were intra-state, but inter-state involvement was a component in both instances. Inter-state involvement was witnessed through international mediation in both sites as well as through other states' involvements in the conflicts. India became militarily involved in Sri Lanka's conflict (for example, Bullion, 1995) while USA–Indonesian relations incorporated aspects of Aceh's situation (for example, Smith, 2003).

Consequently, the type of state involvement as a qualitative typology for disaster-diplomacy case studies adds little more beyond diplomacy track (section 5.2.4) while contributing towards ambiguities. It is useful to classify each party involved as being state or non-state, because that indicates some level of official government sanction or contribution. Classifying the case studies as being state or non-state does not add that much usefulness to the discussion.

5.2.7 *Qualitative typology: active and passive*

Another possibility for qualitatively classifying disaster-diplomacy case studies is demonstrated by the numerous historical battles that were affected by weather. Lewis (1999) notes Robert Louis Stevenson's discussion of the 1889 typhoon in Apia, Samoa that wrecked German, American, and British fleets, preventing a battle. By using first-hand accounts of the weather reports recorded in ships' log books (see Wheeler, 2001), Wheeler (1991) describes the weather during the North Sea Campaign of 1797 that pitted the English against the Dutch (representing the French). Using similar techniques, Wheeler (1995) details the weather leading up to and during the English versus French naval battle of Quiberon Bay on 20 November 1759.

At least three times, Russia's winter weather has severely hindered an invading army, contributing to Russia's victory in the battles and wars. Charles XII of Sweden experienced the Russian winter in 1708–9, Napoleon Bonaparte at the end of 1812, and Adolf Hitler in 1941–43. Ward (1918) details the extensive influence of weather, particularly floods, on battles in Europe in World War I.

In each instance, factors other than the weather significantly influenced the situation and the military results. All these examples are still the inadvertent 'use' of weather in order to influence a military situation. In some cases, violent conflict was avoided, even if inadvertently. In other cases, one side gained a military advantage. These case studies represent the passive 'use' of weather. They provide situations that developed with limited premeditation or intention of the parties involved, even if parties used the weather to their advantage such as the Russians' scorched-earth retreat described later in this section.

That can be contrasted with parts of The Netherlands being deliberately flooded to impede enemy soldiers. Examples occurred in 1672 and by both sides during World War II (Wagret, 1968). This example is active use of disasters to influence conflict. Although the idea was to avoid engagement with the enemy, no claim is made that this is 'disaster diplomacy'. In contrast, it was using floods as a weapon in a war with many other weapons. Similar use of disasters as weapons of war have been documented.

During World War II, both sides considered bombing dams in order to affect their enemy's industrial activity. UK raids were successful in breaching dams in Germany's Ruhr Valley which hurt German industry in the area. Ultimately, that action appears to have had limited impact on the war effort, despite immortalisation and glorification of the bombings in a 1951 book and a 1955 film, both called *The Dam Busters* (Ramsden, 2003).

Earthquakes have been the subject of attempts at control. While earthquake control has long been postulated, unintentional discoveries led to more serious attempts to investigate and harness it. In Arizona in 1935, the artificial reservoir Lake Mead on the Colorado River was discovered to potentially be inducing earthquakes through the weight of the water on fault lines (Smith, 1996; Waltham, 1978). A 1968 underground nuclear-bomb test in Nevada is thought to have induced a moderate earthquake followed by several felt aftershocks (Waltham, 1978).

These experiences led to suggestions of inducing more frequent, lower-magnitude earthquakes with underground explosions in order to prevent a single high-magnitude shock. Alternatively, reservoirs could be filled to induce low-magnitude earthquakes or a reservoir could be drained to prevent a large earthquake.

Reservoirs increasing groundwater pressure leading to reduced friction along a fracture, causing low-magnitude shocks, has also been considered. Such 'lubrication' of a fault could potentially be achieved by injecting water into deep boreholes, a technique discovered in the 1960s near Denver leading to experiments around Colorado (Bolt, 1993). Much of this work is speculative. It carries the risk of creating a major earthquake – or of appearing to have created a major earthquake. While the investigations were ostensibly for reducing earthquake damage, such processes could easily be considered for inducing a major earthquake in an enemy's territory.

Biological warfare, such as attempts to create epidemics, are other forms of weapons of war. These efforts have a long history. Hannibal used snakes to attack an enemy in 184 BC (Noah *et al.*, 2002). The introduction of the plague to Europe leading to the Black Death is postulated to have occurred in 1346 through its use as a weapon at the siege of Caffa, which is now in the Ukraine (Wheelis, 2002).

Weather modification, including redirecting tropical cyclones, has a long history as well. Cotton and Pielke, Sr. (1995), Glantz (2000) and Kahan *et al.* (1995) provide details on experiments and operations for cloud seeding that have been used around the world. They have been applied to attempts at suppressing hail, increasing or directing precipitation, and affecting the tracks and intensities of tropical cyclones.

The success of any of the attempts is continually in dispute. Even where successes are accepted, concerns are raised about the operations. Does the possibility exist of making the situation worse, such as causing a major flood rather than simply increased rainfall? Could litigation occur if damage might be attributable to the weather modification operations – even if less damage occurred than would have been expected without the weather modification?

In 1947, General Electric Corporation in collaboration with the American Army started a project aimed at seeding hurricanes to make them less intense. One hurricane that was seeded suddenly changed direction and made landfall in Georgia and South Carolina, causing damage and angering people who had been affected (Kwa, 2001). Legal concerns interfered with scientific arguments regarding whether or not the seeding had caused or had contributed to the change of the hurricane's track. That experience dampened future efforts to seed hurricanes in the USA.

Efforts were made by the USA's Department of Defense during the Vietnam war to seed clouds over Laos, Vietnam, and Cambodia in order to hamper movement by those fighting the Americans (Shapley, 1974). Glantz (2000) notes that Cuba then became worried that such 'Weather Warfare' (Shapley, 1974: 1059) could be used to try to generate weather damage in Cuba. Possibilities would be storms, hurricanes, or droughts.

The revelations about Vietnam led to multilateral efforts to prohibit 'environmental warfare' (Goldblat, 1975). One result was the 1976 'Convention on the prohibition of military or any hostile use of environmental modification techniques' (now the 'Environmental Modification Convention') that entered into force in 1978.

Additionally, Glantz (2000) describes how many countries affected by tropical cyclones need the water from the storms. Deliberately affecting the tracks or intensity of the tropical cyclones could lead to water-resource problems even if hurricane damage were reduced. Active control of weather, irrespective of beneficial objectives, can lead to detrimental impacts.

The distinction between active and passive use of weather in conflict is not strictly binary. In the cases of the Russian winter, the people retreating from the advancing armies frequently employed 'scorched-earth' tactics, destroying or taking all their assets. That left the invading armies with few food or fuel resources. Combined with the harsh winters, the occupiers were defeated.

The people retreating could not fully know what the winter would be like. They did know that the weather could impede the invaders. They supported the weather by leaving little for the invaders to use against the winter. That is a combination of active and passive use of weather as a weapon of war.

Disaster diplomacy can also be examined from active and passive perspectives (Kelman, 2003). Many case studies (see Table 5.6) describe situations where disaster-related activities and diplomacy interacted without those

Table 5.6 Active and passive disaster diplomacy

	Active or passive disaster diplomacy?
Burma and Cyclone Nargis	Mainly passive, but with some deliberate and unsuccessful attempts to make it active
Casualty identification	Mainly passive
China and the 2008 earthquake	Active in that the parties knew the implications and accepted those
Cuba–USA	Mainly passive, but with strong recognition that disaster diplomacy could occur which led to active attempts to avoid it
Ethiopia–Eritrea	Active
Greece–Turkey	Passive from the perspective of the governments, but the media and people made it active
Hurricane Katrina	Mainly passive
India–Pakistan	Mainly active
Iran–USA	Mainly passive with active attempts to avoid disaster diplomacy
North Korea	Mainly active
Philippines	Mainly passive
Sea-level rise and island evacuation	Mainly active
Southern Africa 1991–93	Mainly passive
Tsunami, Aceh	Mainly active
Tsunami, Sri Lanka	Mainly active
Tsunami, other case studies	Most were passive
Vaccination programmes	Mainly passive, but some authors promoting more active approaches

involved necessarily being aware of the entire extent, or possibilities, of that interaction. They also did not know how to possibly manipulate the interactions for the gains that they might desire. That suggests passive disaster diplomacy.

In other cases (Table 5.6), deliberate attempts were made to invoke disaster diplomacy or to seek connections between the disaster-related activities and the diplomacy. For Eritrea–Ethiopia, this active approach was the impetus behind disaster diplomacy. For Iran–USA after the 2003 Bam earthquake, the role that the disaster could play in the two countries' relations was fully recognised and pushed by many politicians and the media. That potentially led to disaster diplomacy's downfall in this case study.

Kelman (2003) suggests that through active disaster diplomacy, opportunities could be monitored for likely disaster-diplomacy candidates. When a warning system is being set up, when joint monitoring efforts are proposed, or after a disaster strikes a key area, then disaster-diplomacy proponents could leap into action. They would be trying to ensure that the disaster-related activities do indeed support a long-term easing of tensions amongst antagonistic parties.

Active disaster diplomacy may involve working with the media or lobbying governments or other organisations. The message would be that diplomacy can and should be created or should improve due to disaster-related cooperation.

Since disaster-related activities are about saving lives, the rhetoric could extend to disaster-related activities being one of the best forms of bringing enemies together. Especially in post-disaster situations, the humanitarian imperative of saving lives and the compassion evoked when witnessing human suffering should form the basis for demanding disaster diplomacy. Then, support would be provided when hostile parties move towards reconciliation.

This approach to active disaster diplomacy could easily succumb to legitimate accusations of naivety (see section 6.1) and lack of ethics (see section 9.1). The approach could be tempered to evaluate whether or not disaster diplomacy is likely to succeed, especially with or without different forms of intervention. Where intervention is deemed to be appropriate, without naivety or ethical violations, then active disaster diplomacy could be implemented. When it appears unlikely that disaster diplomacy would work, with interventions potentially causing more problems than they solve, then active disaster diplomacy would be (actively) avoided.

5.2.8 Summarising the qualitative typologies

Six disaster-diplomacy qualitative typologies have been presented: propinquity, aid relationship, diplomacy track, disaster-diplomacy purpose, type of state involvement, and degree of active or passive disaster diplomacy (Table 5.7). Limitations of and complexities in these typologies have been explored.

Table 5.7 Summary of disaster-diplomacy typologies and categories

Typology	Categories of the typology	Examples
Propinquity	1 Single geographic entity 2 Physically connected 3 Near but not physically connected 4 Not near	1 Tsunami, Aceh 2 Greece–Turkey 3 Cuba–USA 4 Iran–USA
Aid relationship	1 Mutual aid 2(i) Combined aid: parties in conflict coordinating 2(ii) Combined aid: friendly parties coordinating to another party with conflict 3 Donor-recipient	1 Greece–Turkey 2(i) North Korea 2(ii) Burma, Cyclone Nargis 3 Sea-level rise and island evacuation
Diplomacy track	1 Government 2 Professional conflict resolution 3 Business 4 Private citizens 5 Research, training, and education 6 Activism 7 Religious 8 Funding 9 Public opinion/communication	1 Sea-level rise and island evacuation 2 Tsunami, Aceh 3 Hurricane Katrina 4 Greece–Turkey 5 Cuba–USA 6 Eritrea–Ethiopia 7 Disaster-casualty identification 8 Vaccination programmes 9 Burma, Cyclone Nargis

Table 5.7 (continued)

Typology	Categories of the typology	Examples
Purpose	Explored in detail further in Chapters 6 and 7	
Type of state involvement	1 Inter-state with state involvement 2 Inter-state with non-state involvement 3 Intra-state with state involvement 4 Intra-state with non-state involvement	1 Hurricane Katrina 2 Vaccination programmes 3 Tsunami, Aceh 4 Colombia (section 5.2.2)

The qualitative typology approach assists in explaining some aspects of disaster-diplomacy case studies and disaster-diplomacy patterns. Not all aspects of disaster diplomacy match the explanatory parts of the qualitative typologies. Do these qualitative typologies provide a useful predictive model for disaster diplomacy?

5.3 No predictive model

All the qualitative typologies assist in explaining characteristics of the disaster-diplomacy case studies. None provides a solid way forward for trying to categorise the case studies or for understanding them as a coherent set. Too many exceptions exist for any rules. Rather than a limit of the qualitative typologies, that might result from the broad conceptualisation of disaster diplomacy.

The necessarily broad approach produces many limitations, inconsistencies, and overlaps when trying to create clear categories or when indicating differences and similarities amongst case studies and their categories. It is also feasible that the qualitative attempts do not go far enough. For instance, all the typologies could be taken simultaneously and quantitatively weighted to yield a quantitative index that would suggest the level of disaster diplomacy present. Would such a quantitative index be credible, even if calculations were completed by considering ranges of weightings for each typology?

Other attempts at qualitative approaches for analysing political impacts of disasters yield similar limitations, inconsistencies, and overlaps as seen for disaster diplomacy. Olson and Gawronski (2010) consider only public views of disaster response from governments. They develop a 5C+A framework consisting of six characteristics, the first letters of which provide the framework's name:

- Capabilities, seen as being short-term resources available or accessible.
- Competence, referring to successful use of capabilities, again interpreted as being short term.
- Compassion, interpreted as perceptions of governmental interest in disaster-affected people.
- Correctness, in terms of honest, fair, and transparent aid.

- Credibility, defined in relation to information provided about the disaster.
- Anticipation, which is revealed to be simply governmental disaster-risk reduction.

As with the other qualitative typologies, Olson and Gawronski (2010) do not attempt to weight each of their framework's components.

The 5C+A framework is analysed in the context of several case studies. They provide some helpful insights, but focus on perceptions of governmental action. That covers only a small portion of disaster-related actions and influences. As shown by the strong involvement of non-governmental participants in many of the disaster-diplomacy case studies, relevant conclusions might be difficult without considering a scope beyond governmental influences and perceptions of those.

A portion of the limitations of all the qualitative frameworks emerges due to being overly mechanistic in the qualitative typologies. Exactly as expected, neither disaster-related activities nor diplomacy-related activities behave like linear machines or linear equations.

Many potential outcomes exist for a given set of observations, but no set of observations can be complete. Disaster diplomacy suffers all the inherent problems of predictability that plague coupled sets of non-linear equations where the solutions diverge significantly with tiny variations in initial conditions. As such, quantitative weightings or attempts to generate quantitative approaches to disaster diplomacy would be unlikely to improve the qualitative typologies. Instead, those attempts would likely end up demonstrating the same limitations as the more quantitative studies discussed in section 5.1.

The consequence is that the qualitative typologies, individually or in different combinations, yield little fine-scale predictability in disaster diplomacy. At the larger scale, the general prediction can be made that disaster diplomacy is unlikely to be observed or to be successful. Even when prospects for disaster diplomacy are recognised early on and active attempts are made to effect disaster diplomacy, success might not be possible and is never guaranteed.

Post-tsunami Aceh (section 4.10) had a strong basis of pre-existing conditions that permitted the tsunami to create a space where peace was possible. Due to the pre-existing conditions, disaster diplomacy could not be new. Post-tsunami Sri Lanka and Eritrea–Ethiopia also witnessed possibilities for disaster diplomacy. Even with, or because of, direct efforts to make disaster diplomacy happen, it did not.

As such, the limited explanatory power of the qualitative typologies is not particularly useful for producing a predictive model indicating the circumstances under which various forms of disaster diplomacy will and will not manifest. Much more can be said about the circumstances under which various forms of disaster diplomacy might and might not manifest, but little is certain.

Most pathways for supporting disaster diplomacy are usually open, but are not usually taken. Meanwhile, pathways that hinder disaster diplomacy are actively or passively sought, especially on the time-scale of months or longer.

No pattern emerges from trying to correlate different levels of disaster-diplomacy successes or failures with complex adaptive systems mechanisms, complex adaptive systems properties, propinquity, aid relationship, diplomacy track, disaster-diplomacy purpose, type of state involvement, or degree of active or passive disaster diplomacy. No case study examined has yet satisfactorily answered any of the sets of questions given, including the revised five questions in section 3.3, to indicate that unambiguous disaster diplomacy was present as a major influence on the wide contexts of the case study.

In summary, the appropriate observation is that evidence does not exist that disaster diplomacy fully succeeds or could succeed fully. That permits the main prediction that disaster diplomacy is unlikely to have many or significant successes. In summary, the reason is the complexities of disaster-related activities and diplomacy-related activities along with the high degree of interconnectedness amongst them.

With these complexities and connections, sorting through the interactions and possibilities for influences on disaster-related and diplomacy-related activities is a significant challenge. That is especially the case when seeking connections between the two for and via disaster diplomacy.

Predictive approaches would thus not necessarily be expected. So it is fair that such predictive approaches to disaster diplomacy are not found – at least, not yet. Realism indicates that each case study has its own connections and contexts dictating the ultimate disaster-diplomacy outcome or, as is more often the case, the absence of a disaster-diplomacy outcome.

5.4 Summarising the typologies

Neither complex adaptive systems nor the taxonomic qualitative typologies nor the quantitative approaches provides a satisfactory result for disaster diplomacy in terms of explaining or predicting patterns observed in case studies. This result is neither surprising nor discouraging. Instead, it is part of the reality that must be dealt with when dealing with complex processes such as disasters and diplomacy.

As such, is this science useable and useful for policy and practice? The answer is yes, because it injects realism into the studies that are being done. Theorists and those with a highly academic approach might desire neat science that provides set patterns. Disaster- and diplomacy-related work can rarely provide that. By openly explaining why disaster diplomacy approaches are neither explanatory nor predictive, and by continually examining all sides of the arguments, that science can be used by policy makers and practitioners to move forward in the context of the scientific uncertainties and unknowns.

One consequence is that describing disaster diplomacy or providing a succinct approach to policy and practice recommendations cannot be represented by a single set of universal or rule-based bullet points describing what disaster diplomacy is and is not. Nor can succinct guidance be provided to indicate the actions which should or should not be taken with regards to disaster diplomacy. Nor could a single flowchart be produced that delineates specific nodes with straightforward options and consequences of making certain choices.

Numerous exceptions to any rules will always exist. Choices will continually be confounded with incomplete information and potential outcomes that are contradictory yet feasible given the evidence available. That is not worrying, but is normal when dealing with realistic scenarios and with reality.

Given the disaster-diplomacy theory developed and analysed so far, the best that could be provided for explaining disaster diplomacy in more generic terms is an awareness of the case studies that have been analysed to varying degrees of breadth and depth. These case studies suggest disaster diplomacy as a process involving actions over the long term.

Any framework developed and applied will shift as disaster-diplomacy processes evolve. That includes through further case studies; for example, an earthquake disaster straddling a boundary separating two enemies. That framework will also shift as disaster-diplomacy processes are made to evolve. An example is a cross-border flood mitigation programme that brings together parties that previously were not prone to cooperation.

Any disaster diplomacy portfolio of options, toolkit, model, or explanatory and predictive approach must evolve as new case studies bring new insights. That will support the understanding of disaster diplomacy through increasing experience and practical examples, both contemporary and historical. That will continue to update the science by indicating what is and is not known.

If the desire exists for some form of disaster-diplomacy predictability or, at minimum, direction, then how could more specific disaster-diplomacy predictions be made near the beginning of a case study? Such predictions would need to go beyond the general statement that disaster diplomacy is not likely to be expected. The answer depends on how much control those involved in disaster diplomacy have or seek.

When do they implicitly or explicitly choose to make disaster diplomacy succeed or fail? How and why do they select and avoid certain pathways that are expected to yield the desired disaster-diplomacy outcome (which could be success or failure)? If different parties involved prefer or seek different pathways, how do various balances of power and resources amongst them determine which preference ends up being reality? Pathways pursued for disaster diplomacy's successes and failures are detailed in the next two chapters.

6 Explaining disaster diplomacy's successes

Despite the overall failure generally observed with disaster diplomacy, many characteristics of case studies reveal some forms of success. In parallel, many parties involved in disaster-diplomacy situations are hoping for success. They work actively towards that.

This chapter discusses specific pathways that are pursued to seek successful disaster-diplomacy outcomes. The discussion includes two forms of disaster diplomacy that display positive results: tit-for-tat disaster diplomacy and mirror disaster diplomacy.

6.1 Success pathways

Even though the humanitarian imperative infrequently dominates long-term diplomatic activities, aspects of disaster diplomacy can nevertheless be present. Disaster diplomacy is a process occurring alongside, and interacting with, other processes. Examples range from electioneering to war and from natural resource extraction to sustainable development (however that is defined). Even where disaster diplomacy is actively opposed by some involved in the process, others might be passively opposed, neutral, passively supportive, or actively supportive.

For Greece–Turkey disaster diplomacy, some politicians were reluctant to speed up the ongoing backroom diplomacy, but they were dragged along by the people and the media (Ker-Lindsay, 2000). For Cuba–USA disaster diplomacy, both governments tended to object to disaster diplomacy in words and actions, but it is pursued by scientists and technical organisations (Glantz, 2000). India–Pakistan disaster diplomacy has been led by government leaders, generally garnering popular support for their actions, but fighting against some vociferous opposition. Ultimately, they have been finding more diplomatic success outside of disaster-related activities.

Therefore, to implement disaster diplomacy actively requires the long-term patience, creativity, and flexibility required of other political processes. Due to these inherent complexities, a robust framework of action and specific action recommendations would be challenging to formulate and defend. Even if a specific theoretical framework were generated, it might be inappropriate to apply it in reality due to the number of exceptions, provisos, and instances that do not fit the pattern.

Instead, a set of disaster-diplomacy possibilities, based on the experiences from case studies and the hypothesising beyond the case studies, does exist. This set forms a disaster-diplomacy portfolio, toolkit, or repertoire from which ideas, actions, and tools could be selected. That selection yields a framework of action which would be specific to each situation and to the interests of those who are involved in disaster diplomacy.

The portfolio, toolkit, or repertoire is provided as pathways which promote disaster diplomacy (Table 6.1) and pathways which oppose disaster diplomacy (Chapter 7). The purpose of these pathways is to demonstrate the approaches available to those involved in disaster diplomacy. These parties are trying to determine whether or not to try to influence disaster diplomacy – and, if so, how.

If the choice were to influence disaster diplomacy actively, then the party involved would need to determine whether to try to promote or inhibit disaster diplomacy. Then, the lists of pathways would provide the possible directions to follow.

Table 6.1 Pathways promoting disaster diplomacy

Pathway name	Case-study examples
Avoid forcing	China and the May 2008 earthquake
	Ethiopia–Eritrea
	Greece–Turkey
	India–Pakistan
	North Korea
	Tsunami, Aceh
Focusing on disaster, not diplomacy	Cuba–USA
	Iran–USA
	Southern Africa, 1991–93
Informal networks	Cuba–USA
	Greece–Turkey
Multiple levels/tracks	Greece–Turkey
	India–Pakistan
	Sea-level rise and island evacuation
Multi-way process	Greece–Turkey
	India–Pakistan
	North Korea
	Tsunami, Aceh
Science	Cuba–USA
	Middle East seismology, e.g. Middle East Regional Cooperation Program: www.relemr-merc.org and Middle East Seismological Forum: www.meseisforum.net
Symbolism	China and the May 2008 earthquake
	Hurricane Katrina
	India–Pakistan

The choice might be partially directed by the similarity or differences of the situation being analysed to the case studies in which each pathway had previously been used. Even if a pathway is used, that does not mean success as measured by the party's intention in using that pathway.

As indicated by the case studies of disaster diplomacy, not all pathways were attempted and not all pathways were successful in the case studies in which the pathways were applied. As emerges from the discussion on disaster diplomacy's failings, no one could assume that the choice made or that the pathways selected would yield the desired outcome. That is especially the case because different parties involved often have different desired outcomes.

Meanwhile, diplomatic pathways other than disaster diplomacy exist. For example, talking for the purpose of reconciliation rather than for any other reason. Those pathways are not included in this discussion. The pathways presented here are specifically related to disaster diplomacy.

'Avoid forcing' refers to the fact that disaster diplomacy cannot be forced nor can it be presumed to work. As with most diplomatic processes, care and extensive communication are needed over the long term to build trust and to avoid misunderstandings or missteps. Additionally, a pace is needed which is not so fast that decision makers lose track of events and decisions, but not so slow that progress is limited or is overtaken by other events.

'Focusing on disaster, not diplomacy', means declining to link disaster-related and diplomatic activities. Such an approach obviously inhibits disaster diplomacy in a direct manner. Conversely, and more importantly, that approach might permit disaster-related cooperation to lay the groundwork for later diplomacy. Collaboration on disaster-related activities without further expectations can build trust, make connections, and illustrate successes. Enemies could then use that, if they choose, to generate confidence for pursuing other collaborative endeavours.

'Informal networks' refer to communication and interaction away from formal settings. They can be effective in providing disaster diplomacy opportunities and in laying the groundwork for formal negotiations. Examples are scientists sharing real-time data on environmental phenomena such as volcanic eruptions or insect invasions along with people donating money and goods to a disaster-affected enemy. Additionally, disaster-related organisations, non-governmental or governmental, from parties in conflict could meet in a neutral location or in international settings in order to share their ideas and experiences.

'Multiple levels/tracks' as a pathway means that disaster diplomacy along a single diplomacy track is usually unsuccessful. If disaster diplomacy is led only by sovereign governments, then people can undermine the efforts. In November 2005, Sri Lankans elected a President who vowed to take a hard line against the Tamil Tigers. That was despite the (ever-diminishing) possibility for post-tsunami resolution of the violent conflict (section 4.10). The President took the promised hard line and soon ended the violent conflict through military means.

If disaster diplomacy is pursued by only certain diplomatic tracks or organisational approaches such as activism, the media, business aims, or religious interests, then the process could have minimal credibility with other tracks, organisational approaches or governments. If disaster diplomacy is initiated and advocated at only the grassroots level, then 'While public opinion is undoubtedly powerful, it is also fickle ... what the people have given directly, the people can take away' (Ker-Lindsay, 2000: 216, 229). The 'multiple levels/tracks' pathway for disaster diplomacy indicates that a combination of parties across multiple diplomacy tracks is needed for disaster diplomacy to succeed. That is exactly the point of multi-track diplomacy (Diamond and McDonald, 1993).

Requiring a 'multi-way process' indicates that, without exchanges amongst all those involved in disaster diplomacy, it is unlikely to proceed far. Rather than one state, organisation, or diplomacy track pushing the process, making concessions, or suggesting how to proceed, a continual exchange of ideas and steps is needed to achieve the desired success.

Even in donor–recipient case studies, the donor and the recipient need to be involved in a multi-way process. Statements from the recipient would be needed for disaster diplomacy to continue, such as public gratefulness to the donor. Subsequent actions would also be needed, such as a deliberate reduction of hostilities or such as permitting easy access for goods and people that are helping. Simultaneously, statements and actions from the donor would be needed for disaster diplomacy to continue. Statements should avoid schadenfreude while subsequent actions could be deliberate reduction of hostilities or vindictiveness.

The pathway of 'science' is related to some degree to informal networks. Scientific and technical exchanges at all governmental and non-governmental levels, in addition to individuals taking initiatives, have frequently provided a powerful basis for more cooperation.

Scientists often articulate that they believe themselves to be apolitical and immune to concerns and subjectivities of politics, including diplomacy. Despite this naivety – Martin (1979) debunks the myth of the objective scientist – claiming to be apart from politics and claiming to be neutrally objective can sometimes be helpful for enabling disaster-related interaction amongst enemies.

As Glantz (2000) notes, Cubans and Americans have been involved in many scientific exchanges and try to divorce their governments' politics from the scientific collaboration. Petropoulos (2001) suggests that the Greece–Turkey thaw at that time should be providing an opportunity for scientists from both countries to collaborate further on disaster research.

'Symbolism' is the final pathway in Table 6.1. It refers to the use of disaster-related activities for making diplomatic points. Symbolism which promotes disaster diplomacy occurs when recipient states (i) do not necessarily need proffered assistance but accept it or (ii) become donors, even when that is not necessary.

Symbolism can backfire. Cynics might suggest that when those who are traditionally recipients decide to donate, they are ingratiating themselves to a major usual donor in order to extract benefits later. Alternatively, they might be attempting to demonstrate that they deserve to play a more prominent role on the international stage because they can be a donor.

Symbolism of that nature can be used to reject disaster-related collaboration, thereby precluding disaster diplomacy. One reason for India initially rejecting aid after the 26 December 2004 tsunamis and the 8 October 2005 earthquake was to take the role of being a donor, rather than being a recipient. India had accepted international aid after the 26 January 2001 earthquake. India then explained that the scale of the 26 December 2004 tsunamis in India was far less than the 2001 earthquake. After the 2004 tsunamis, India provided aid to the Maldives and to Sri Lanka.

It would be unfair to suggest that Indians suffered because their government rejected aid for purely political and pride-related reasons. Instead, India used the disasters to demonstrate that it has the capability for taking care of its own disaster relief because India can do so and because India was hoping to gain respect internationally. India has extensive domestic and international experience and capability in post-disaster assistance along with disaster-risk reduction (for example, AIDMI, 2007; SEEDS, 2007). India trying to refuse assistance was not due to only symbolism, even though symbolism played a role in that.

6.2 Further success: tit-for-tat

Tit-for-tat disaster diplomacy is the possibility that one party providing aid to its enemy could produce a similar and reciprocal gesture in the future, despite continuing conflict between the parties. In parallel, tit-for-tat could be the refusal of one party to aid another party during times of need which could lead to a reciprocal refusal to assist in the future, creating or perpetuating conflict amongst the parties.

Tit-for-tat gestures linked Hurricane Katrina disaster diplomacy in 2005 to the tsunami diplomacy after the 26 December 2004 Indian Ocean tsunamis. Several states that had been hit by the tsunami, and that had received aid from the USA, offered post-Katrina assistance to the USA. They commented that they remembered the generosity from Americans and from the US government, so they wished to return the favour in the USA's time of need. Sri Lanka, for example, called its $25,000 donation following Katrina 'a token contribution'. Bangladesh offered $1 million to the USA, which was accepted, along with disaster-response expertise, which was not accepted.

Despite Iran electing a President unfriendly towards the USA on 24 June 2005, Iran offered post-Katrina assistance to the USA. Strong allusions were not made to the USA's previous aid to Iran after earthquakes, most notably in 2003 after the Bam earthquake. That leaves it as an open question regarding the tit-for-tat traits of the offer.

Iran's offer of disaster assistance also came alongside the US government's refusal to provide visas for some Iranian officials who were applying in order to attend the 7–9 September 2005 UN World Summit in New York. Regarding Katrina, Iran respected diplomatic niceties, with Iran's Foreign Ministry stating, 'If Iran's help is needed and requested we'd respond to the call; however we should make sure first that there is such a request'. Iran was also straightforward in denying any political connections to the aid. That emulated the Iran–USA interaction regarding aid offers from the USA to Iran after the Bam earthquake. To some extent, the efforts to disconnect disaster aid from diplomacy for both Katrina and Bam could be viewed as being tit-for-tat in terms of the deliberate avoidance of disaster diplomacy.

Another case study displaying tit-for-tat characteristics is Greece–Turkey (Ker-Lindsay, 2000, 2007). After the September 1999 earthquake in Athens, many offers of assistance from Turkey – from the grassroots up to the ministerial level – specifically noted that it was a reciprocal gesture, following Greece's aid to Turkey after the August 1999 earthquake. Many Turks were deliberate in offering, as a sign of respect, assistance that imitated what Greece or certain Greeks had provided to Turkey a few weeks before.

Weaknesses in tit-for-tat disaster diplomacy are more evident than the apparent strengths. North Korea's famines provide an example. Starting in 1995, the USA had provided aid to North Korea and was involved in complex negotiations over numerous topics. Some press reports suggested that North Korea had offered rescue assistance to the USA after Hurricane Katrina, but at the time, USDS (2005f) did not list North Korea as a potential donor.

As Hurricane Katrina's impacts were becoming evident, the six countries involved in negotiating with North Korea (China, Japan, North Korea, Russia, South Korea, and the USA) reached an agreement regarding North Korea's nuclear ambitions and development assistance to the country. That agreement ended talks that had lasted three years. The conclusion of the talks was not linked to Hurricane Katrina.

East Timor provides another example of failed tit-for-tat disaster diplomacy regarding Hurricane Katrina. On East Timor's independence day of 20 May 2002, President Xanana Gusmao expressed gratitude for the USA's role in reaching that day. East Timor was not listed by USDS (2005f) at the time as being amongst the states that offered post-Katrina assistance.

States lower on the Human Development Index (UNDP, 2005) than East Timor were listed, including Guinea, Togo, and Yemen. No claim is made that East Timor should have offered aid nor does any implication exist that East Timor was insensitive to the USA's suffering. The lack of tit-for-tat disaster diplomacy could have been from Gusmao identifying Bill Clinton as being the main American supporter of East Timor, rather than focusing on the US government or the American people more generally. The lack of tit-for-tat disaster diplomacy could also have been because East Timor recognised that the USA was obtaining needed aid from other sources and that East

Timor could add little. East Timor was still, fairly, focused on its own state building.

Tit-for-tat disaster diplomacy has a further limitation in that some disaster-diplomacy case studies progress or fail to progress through mutual agreement rather than through unilateral decisions by key parties in an ordered sequence. Despite the tit-for-tat disaster diplomacy characteristics, Greece–Turkey rapprochement experienced more successes from the complex, pre-disaster, intergovernmental negotiations over many topics of mutual interest (Ker-Lindsay, 2000, 2007). India–Pakistan disaster diplomacy after the 2001 Gujarat earthquake appears to have started from a unilateral initiative from Pakistan, but then evolved according to the Indian and Pakistani governments' mutual interests and disinterests in resolving differences (Kelman, 2003).

Tit-for-tat disaster diplomacy is an element that contributes to disaster diplomacy, but it does not always succeed or have the largest influence. It should not be relied on, and should not necessarily be encouraged, for long-term disaster-related successes.

6.3 Further success: mirror disaster diplomacy

Regarding Cuba–USA relations in the context of disaster preparedness, Naranjo Diaz (2003: 62–63) wrote:

> Cubans have been forced to be more efficient in facing natural disasters in a scenario of political conflict with the US government. This is maybe an opposite view of the disaster diplomacy approach. Protective measures under a conflict are developed in such a way that the enemy would not be able to take advantages from the disaster.

This idea of 'mirror disaster diplomacy' might be more prominent than disaster diplomacy because it depends on internal policies and choices rather than on bilateral or multilateral negotiations. The political will for self-sufficiency, or to avoid appearing vulnerable, could be a strong enough impetus to encourage a political entity to invest so much in disaster-risk reduction that less external assistance is required for disaster-related activities.

Nevertheless, a large catastrophe, for example at the national scale, might still require international assistance. Despite mirror disaster diplomacy, Cuba requested aid following Hurricane Michelle in 2001 (Kelman, 2003) and Hurricane Dennis in 2005 (Kelman, 2006a). Similarly, North Korea had prided itself on self-proclaimed self-sufficiency and had limited its overt contact beyond its international borders, collaborating mainly with China and Russia. The 1995 floods, droughts, and famines were severe enough to force North Korea to the international negotiating table, involving Japan, South Korea, and the USA in addition to China and Russia.

Consequently, mirror disaster diplomacy has a limit to which it creates or could create disaster-related self-sufficiency. This limit is partially practical.

Disasters can be large enough to overwhelm even a rich state's resource base, as seen for Hurricane Katrina affecting the USA.

This limit is partially an active decision as well. The initially estimated damage of $125 billion from Hurricane Katrina (Munich Re, 2006) was under one third of the American Department of Defense's annual budget at that time of over US$400 billion (OMB, 2005). The USA's Department of Homeland Security's annual budget at that time was US$30 billion (OMB, 2005). That should have been sufficient to have prepared for a disaster the size of Katrina without needing supplies and support from outside of the country.

Overall, at the time of Katrina, the USA had the resources and capabilities to be self-sufficient with regards to disaster-related activities. The government had chosen to allocate those resources to other priorities. The focus regarding disasters tended to be counter-terrorism and crisis management at the expense of non-terrorism disasters and pre-disaster actions (for example, Schneider, 2005; Somers, 2005).

As well, during their respective disasters, both the USA and North Korea initially attempted to avoid international assistance despite the mounting evidence that the disasters could not be dealt with domestically. Isolationist attitudes perhaps created a false sense of hope or a false sense of security that the country would be self-sufficient under all circumstances. Furthermore, isolationism might develop a need or desire to avoid (i) perceived loss of face through asking for or accepting help and (ii) dependency on non-domestic parties.

An ethical concern exists with mirror disaster diplomacy. The political will and resources could be found to aim for disaster-risk reduction due to an external political threat, rather than due to a genuine desire to save lives and to avert destruction.

Meanwhile, a practical concern exists that aiming for mirror disaster diplomacy does not always lead to its realisation. The result could worsen disasters if no provisions are made to call for and deal with external assistance. Nonetheless, if mirror disaster diplomacy leads to disaster-risk reduction, then some positive results could emerge, even if the disaster risk reduction occurred to avoid diplomacy.

Mirror disaster diplomacy is another element that must be considered in disaster diplomacy. It does not always succeed and it can potentially worsen disaster-risk reduction while negatively impacting diplomacy. It should not be relied on, and should not necessarily be encouraged, for long-term disaster-related successes.

7 Explaining disaster diplomacy's failures

Contrasting with the pathways for disaster diplomacy's successes (Chapter 6), many pathways exist and are pursued to try to make disaster diplomacy fail. Some emerged in Chapter 6's discussions. This chapter discusses specific pathways that are pursued to inhibit disaster-diplomacy outcomes, including forms of disaster diplomacy that indicate how disaster-related activities can exacerbate conflict.

7.1 Failure pathways

The pathways which inhibit disaster diplomacy are listed in Table 7.1.

'Avoiding diplomacy' or the appearance of diplomacy is a reason for refusing to collaborate on disaster-related activities. The concern of the parties involved might not be specifically to avoid disaster diplomacy, but to avoid any form of diplomacy. Alternatively, this pathway can be invoked to demonstrate independence.

Nelson (2010a) analysed a dataset of instances from 1982 to 2006 where disaster aid was refused. He concluded that states that had recently undergone a major governance transition, such as gaining independence, were more likely to decline disaster aid, claiming to demonstrate internal capability in dealing with a disaster. No diplomacy was desired, so the states felt it appropriate to decline aid, whether or not that aid was needed.

In fact, relying on disaster-related activities to advance diplomacy; that is, 'dependency on disaster', can be counterproductive. Many disaster-related activities occur over the short term, especially relief and response, whereas diplomacy usually requires extensive efforts over the long term. Little expectation exists that depending solely on disasters would be likely to produce lasting diplomatic success, although the wider scope of disaster-related activities over the long term might have more chances for success.

Disasters can be used as weapons, as detailed in section 5.2.7. Employing this pathway could make it difficult for parties in conflict to reconcile over this, or a similar or parallel, disaster. During the 1998 drought, Cuba requested international assistance but refused aid from the USA claiming that the USA's trade embargo against Cuba was one factor in Cuba's inability to

Table 7.1 Pathways inhibiting disaster diplomacy

Pathway name	Examples
Avoiding diplomacy	Burma and Cyclone Nargis Cuba–USA Ethiopia–Eritrea Iran–USA and including Iran's refusal to accept aid from Israel
Dependency on disaster	This pathway is evident to some extent for India–Pakistan and North Korea, but the evidence does not suggest that these diplomatic processes are wholly dependent on disaster
Disaster as a weapon	Several examples exist throughout history (see sections 5.2.7 and 7.1, weather modification and disasters as weapons of war)
Disasters worsening relations	India–Pakistan Tsunami, Sri Lanka
Distraction	Ethiopia–Eritrea
Events overwhelming disasters	This pathway is evident to some extent for Greece–Turkey Cuba–USA Hurricane Katrina North Korea Iran–USA
Expectations	Burma and Cyclone Nargis China and the May 2008 earthquake Greece–Turkey India–Pakistan Iran–USA
False propinquity	Several case studies display this pathway to some extent, but none have enough evidence to support the claim that it affected disaster diplomacy
Spotlight	Burma and Cyclone Nargis Cuba–USA Greece–Turkey India–Pakistan Iran–USA Tsunami, Aceh Tsunami, Sri Lanka
Vindictiveness	Accusations were made that this pathway was chosen for: • Hurricane Katrina, by Cuba, Iran, and Venezuela • USA when offering help to Cuba • Eritrea when offering to help Ethiopia

deal with the drought on its own (Glantz, 2000). Cuba was not blaming the USA for the drought, but did implicate the USA in the drought disaster; that is, in Cuba's need for aid.

Disasters frequently worsen relations amongst enemies. That particularly occurs when the disaster does not involve an environmental phenomenon and when an enemy is considered to have been involved in the disaster.

India's parliament was attacked by militants on 13 December 2001. Although Pakistan's government along with Kashmiri separatists condemned the attacks, India–Pakistan relations suffered because India accused Pakistan of being involved in the violence. Another example is Lewis (1999) describing how the 1970 cyclone that struck East Pakistan, followed by mismanagement of the disaster, provided a trigger for the war that led to the formation of Bangladesh (section 1.2). The cyclone disaster and dealing with it provoked a simmering situation, leading to violent conflict. Nelson (2010b) provides several other cases of disasters involving environmental phenomena leading to conflict.

The 'distraction' pathway means that disaster diplomacy is not always an appropriate pathway for disaster-related activities or for diplomacy. Even if those involved in either disaster-related activities or diplomacy support disaster diplomacy, the process might detract from core issues and long-term solutions regarding both disaster-related and diplomatic activities. Both disaster-risk reduction and diplomacy are long-term endeavours which might not be addressed properly if those involved are distracted by short-term or superficial hype or hyperbole surrounding disaster diplomacy.

The pathway of 'events overwhelming disasters' suggests that non-disaster processes or events can influence diplomacy more than disasters. The Elián González incident of 1999–2000 affected relations between Cuba and the USA more than hurricane preparedness and response.

González was a Cuban boy taken illegally to Florida by his mother via boat. She died on the voyage. González's relatives in Florida wanted him to remain in the USA. González's father in Cuba wanted González returned to Cuba. Court decisions in the USA ('Elián González ... ', 2000; 'Lazaro Gonzalez ... ', 2000) were required to repatriate González.

Another example of events overwhelming disaster was George W. Bush's election to the US Presidency in November 2000 followed by his January 2002 State of the Union address. The speech labelled North Korea and Iran as being part of the 'Axis of Evil' (Bush, 2002). That hindered US relations with those two countries despite disaster-diplomacy possibilities.

The 'expectations' pathway is about raising diplomatic expectations beyond what could realistically be achieved. That can hamper rapprochement when the parties involved become impatient or are seen to have failed to achieve the expectations.

Expressing expectations that disaster-related activities should result in peace could apply undue pressure to diplomats, politicians, the media, or the people to achieve results, irrespective of whether or not they are ready or willing to seek rapprochement and to complete the details. That could potentially force results within a narrow window of opportunity, forming an impetus to push actively for disaster diplomacy to succeed swiftly. If immediate expectations are not met, then the response could be blaming the other side, reinstating and reinforcing previous enmity. That was the case during the India–Pakistan summit of July 2001 that collapsed after having been set up in the wake of Pakistan's post-earthquake offer of assistance to India in January 2001.

'False propinquity' refers to the assumption that states being near each other should cooperate simply on the basis of their propinquity. In reality, propinquity's role in promoting a humanitarian imperative alongside rapprochement could be overstated. Trying too much to be too friendly in the short term, just because of proximity and despite vast cultural and ideological distances, could jeopardise longer-term, genuine, lasting reconciliation. This criticism is tempered by the Greece–Turkey and India–Pakistan case studies using propinquity as one basis for disaster diplomacy. In contrast, Cuba–USA and Eritrea–Ethiopia have not been helped by propinquity.

Disaster diplomacy can further be inhibited by the 'spotlight' pathway. When a peace process becomes prominent, especially following a disaster, a spotlight is often placed on every word, action, and innuendo which is publicised, analysed, and misanalysed. That generates news, satiates public interest, and has high political value, including as a target for those opposing any peace process.

After the 2003 Bam, Iran earthquake, US Secretary of State Colin Powell made remarks that were interpreted as softening the US government's stance towards Iran. Powell was simply repeating the US government's view of Iran that had been outlined two months earlier by his Deputy. The lack of change was conveniently overlooked in the euphoria of promoting Iran–USA earthquake diplomacy. A spotlight on a peace process can also make it an easy target for critics, particularly when that peace process is based on, or is presumed to be based on, as tenuous a connection as disaster-related activities (Ker-Lindsay, 2000).

The final pathway from Table 7.1 is 'vindictiveness'. Offering aid can be used to take advantage of an enemy's troubles or to highlight an enemy's weaknesses in contrast to one's own strengths. The goal would be humiliating the enemy, perpetuating or augmenting the conflict, or seeking revenge. Even disaster-risk reduction knowledge, skills, and techniques could be offered in order to appear friendly while implying that an enemy is incompetent in taking care of its own people.

As with the pathways which promote disaster diplomacy (section 6.1), pathways which inhibit disaster diplomacy cannot be prioritised in terms of relative importance or relative impact. Instead, they form a portfolio, toolkit, or repertoire of possible approaches. Some would be more important and influential, and some would be less important and influential, depending on the specific situation and the specific parties involved.

Each pathway promoting and inhibiting disaster diplomacy is relevant to case study categories across a variety of the qualitative characteristics discussed in section 5.2. That further affirms the limited explanatory and predictive powers of the typologies presented.

7.2 Further failure: inverse disaster diplomacy

Inverse disaster diplomacy refers to diplomatic decisions directly harming domestic disaster-risk reduction. On 4 May 2007, tornadoes ripped through

Greensburg, Kansas. Aside from widespread destruction of infrastructure, one highlighted consequence was the heavy destruction of larger vehicles belonging to the municipality that were needed for disaster response.

In such instances, the state Governor has the right to call out the state's National Guard to assist. Kansas's Governor complained that she was limited in that option due to the large deployments of National Guard personnel and equipment to the ongoing wars in Iraq and Afghanistan. Guard personnel were eventually in place approximately 36 hours after the tornadoes with significant amounts of larger vehicles arriving almost 12 hours after that.

In the months before this disaster, Kansas's Governor along with several other state governors had sought action to try to improve the situation of the National Guard with regards to domestic capability to deal with disasters. They pointed out that the equipment and people overseas left their domestic resources depleted. They were not convinced that they would be able to respond fully to a state or national emergency. Those fears proved to be well founded in Greensburg.

While trying to deal with the situation, the US government and Kansas's government ended up in a war of words. Kansas's Governor at the time was a Democrat while the US President was a Republican. When Kansas's Governor complained about the lack of resources for dealing with the disaster, the White House responded that the Governor was at fault for not requesting the needed assistance. The White House later admitted that the Governor had actually made several requests for help, implying that she had followed proper procedures.

Lewis (2007) suggested such as situation as being 'inverse disaster diplomacy'. With the USA involved in conflict-related diplomacy through the overseas wars, US states and the national government started bickering over the responsibility for responding to the Greensburg situation and the lack of resources for doing so. Lewis (2007) suggested that 'Disaster diplomacy, like its other forms, can sometimes have negative, not just nil, results'.

The White House's attempt to deflect criticism and to make the post-disaster situation politically partisan reflects that form of negative results. Greensburg and the government of Kansas were dealing with an emergency situation and were trying to use state resources for that. Due to federal decisions, the resources were in other countries and were not available to assist domestically. When federal resources were then requested, the situation led to a conflict of accusations and counter-accusations rather than to collaboration for addressing the disaster concerns. 'Inverse disaster diplomacy' manifested.

7.3 Further failure: disaster-related activities exacerbating conflict

As discussed as part of the disaster-diplomacy hypothesis (section 3.2), disaster diplomacy fails in numerous instances through disaster-related activities exacerbating conflict. It can also fail by providing an opportunity to reiterate parties' concerns about the others and by providing the space to ramp up tension.

Le Billon and Waizenegger (2007) even go so far as to suggest that 'Most studies suggest that 'natural' disasters exacerbate pre-existing conflicts' (p. 411). That is an exaggeration. As they reference, plenty of studies do indeed reach that conclusion and those analyses are robust.

Many other studies, such as those referenced in this book, indicate the opposite. As per disaster diplomacy's conclusions, many studies show instead that disasters are frequently neutral with respect to pre-existing conflicts over the long-term. Mandel (2002), for instance, gives mixed results regarding disasters exacerbating conflict when examining Japan's 1995 earthquake, Montserrat's volcanic eruption since 1995, Poland's 1997 flood, Hurricane Mitch in Honduras in 1998, Turkey's earthquake in 1999, and Mozambique's flood in 1999.

Nelson (2010b) specifically examined how disasters could create interstate conflict by examining a dataset from 1950 to 2006. He could not find any examples of a disaster being used as an opportunity for a rival to attack. He did show cases where recent political violence increased the likelihood that a disaster would lead to new conflict. Again, the findings are balanced amongst different conflict-related impacts of disasters.

Meanwhile, Caroll *et al.* (2006) investigated six communities in the USA that experienced wildfire, showing how conflict between locals and non-locals increased due to the disasters. They also explicitly state that many instances of disaster-related cooperation were witnessed. Their conclusions are useful that, at the local level, the same wildfire can both foment conflict and support cooperation in the same location.

A useful pattern from the literature is that many of the quantitative studies (section 5.1) conclude that disasters are more likely to worsen ongoing conflicts than to have no effect or to lead to peace. As discussed in section 5.1, these conclusions are only as good as the data. The data display significant limitations. The studies are not necessarily bad science, because they might be the best feasible given the data limitations.

One particular concern is often the explicit focus on only armed conflict, for which definitions can be ambiguous, despite solid attempts to seek clear definitions (for example, Farer, 1971; Strand *et al.*, 2004). That does not invalidate the studies analysing the data, because definitions are needed to move forward with research. That does suggest the possibility of conclusions being an artefact of the sample selected. It also encourages scope for wider investigations that encompass, compare, and contrast different forms and levels of conflict. Running the same analyses with different definitions and data sets would indicate the sensitivity of the quantitative results to the data selected.

Furthermore, jumping to the conclusion that disasters will inevitably, or tend to, exacerbate conflict is premature because the vast majority of these studies consider only specific disaster events or situations, not all disaster-related activities. Akcinaroglu *et al.* (2008, 2011) discuss aspects of a disaster being a 'shock' that can be a push or a step change in the conflict dynamics. Most disaster-related activities do not have that suddenness factor, which Akcinaroglu *et al.* (2008, 2011) attribute to events such as hurricanes and earthquakes.

It would be useful to compare more thoroughly cases with and without the suddenness factor. It would also be useful to examine the suddenness factor within the understanding that some hazards are rapid onset, but most disasters are slow onset (Lewis, 1988; sections 3.1 and 9.2).

Finally, much of the literature examining disaster-conflict interactions does not always clearly separate amongst (i) specific disasters exacerbating existing conflicts, (ii) not changing the level of existing conflicts, and (iii) creating new conflicts. None of this discussion denies that disaster-related activities are seen to detrimentally affect conflicts. In fact, all categories of disaster-conflict interactions have been witnessed and represent failures of disaster diplomacy. To make the study of disaster-conflict interaction more applicable, more nuancing of the case studies would be helpful, especially openly presenting the limitations and different arguments.

Disaster-related activities frequently do not change the level of existing conflicts, as is concluded in several disaster-diplomacy case studies. Eritrea–Ethiopia (section 4.8) is an example in that attempts to deal with the droughts had little impact on the war – at least, from a qualitative perspective, noting that quantitative indices regarding the level of conflict were not examined.

Autesserre (2002) described a different form of case study with a similar result. She analysed the USA's humanitarian food aid to Sudan during the civil war. Her conclusion was that the food aid was a useful instrument in maintaining the status quo of the conflict, which served US government interests. Rather than failing to change the conflict's level, the disaster measure in the form of relief supplies was a factor in keeping the conflict's level without change.

A third example of disasters not changing the level of existing conflict is Taiwan's 1999 earthquake (section 4.13). The aid sent to the island provided an opportunity for Beijing and Taiwan to continue with their normal diplomatic claims and struggle for control regarding Taiwan being independent or part of the People's Republic of China.

In terms of other disaster-related activities, Cuba–USA (section 4.6) illustrates how hurricane preparedness and monitoring can be neutral in terms of affecting ongoing conflicts. Several instances of vaccine-related ceasefires are also described to have had negligible impact on the conflict, neither reducing nor exacerbating it (section 4.16).

Disaster-related activities creating new conflict has less evidence and runs into more definitional ambiguities. Section 1.2 described case studies of the 1970 cyclone in East Pakistan triggering Bangladesh's War of Independence, the 1972 Managua earthquake providing a focus for the Sandinista uprising, and the earthquake in Sparta sparking a slave revolt. The language, based on the references given, implicates the disaster in creating the new conflict – almost. In each of these cases, the conflict was simmering before the disaster. The disaster provided an identifiable turning point in the conflict, but did not create the conflict.

Similarly, the disaster of the 11 September 2001 terrorist attacks in the north-eastern USA were the impetus for two wars. As noted in section 4.9, a coalition under the auspices of NATO but with UN backing attacked Afghanistan on 7 October 2001. The aims were to topple the Taleban regime and to flush out the terrorists protected and supported by the Taleban and assumed to have carried out the attacks. Then, on 20 March 2003, a coalition led by the USA invaded Iraq to get rid of its leader Saddam Hussein. Yet these two conflicts were not rooted in the 11 September 2001 terrorist attacks.

The 1990s witnessed many violent incidents between al-Qaeda and the USA. On 26 February 1993, al-Qaeda tried to destroy the World Trade Center with a car bomb. On 7 August 1998, al-Qaeda killed over 200 people with bombs at the US embassies in Nairobi and Dar es Salaam. The latter atrocities led to the USA firing cruise missiles at alleged al-Qaeda-linked targets in Afghanistan and Sudan on 20 August 1998.

Regarding Iraq, Saddam Hussein was a USA- and UK-created and supported dictator until 2 August 1990 when his military invaded Kuwait. Under a UN resolution, an international coalition counterattacked on 17 January 1991, driving Iraqi forces out of Kuwait, but stopping from a full-scale invasion of Iraq. The conflict between Iraq and several Western countries who were often, but not always, backed up by UN resolutions, then continued until the 2003 invasion. While the Iraq and Afghanistan wars resulted directly from the 11 September 2001 disaster, the conflicts did not. It is a definitional discussion regarding whether or not these two conflicts were created by a disaster of terrorist attacks.

As with the discussion on the disaster-diplomacy question 'Did new diplomacy emerge?' (section 3.3), it can be difficult to determine genuine 'newness' for diplomacy as well as for conflict. It is clear that, in the same way that disaster-related activities can catalyse diplomacy, the above examples display disasters catalysing conflict. The disaster-diplomacy hypothesis (section 3.2) extends to the opposite statement: disasters can catalyse, but not create, conflict, even turning low-key discontentment into a violent situation, such as in Sparta.

Several elements of the disaster-diplomacy hypothesis remain to be investigated for this disaster-conflict hypothesis. Does the disaster-conflict hypothesis apply to disaster-related activities, not just to disasters? That is, do disaster-related activities catalyse, but not create, conflict? For aspects of climate change (section 4.14), the answer appears to be yes.

Similarly, do the time-scale characteristics of the disaster-diplomacy hypothesis apply to the disaster-conflict hypothesis? That is, would disasters or disaster-related activities only sometimes catalyse conflict in the short term, but not over the long term? If so, is the reason because non-disaster related factors would supersede disaster-related factors?

The answer to these questions would seem to be 'yes', as evidenced by the numerous disasters with the potential to catalyse conflict, but which do not.

Examples are (i) the Haiti earthquake of 12 January 2010 which struck during a volatile political time in the country; (ii) Hurricane Mitch in Central America in October–November 1998 affecting numerous countries still hurting from the previous decades of conflict related to the Cold War; and (iii) the South American cholera epidemic which started in 1991 in Peru, but which appears to have had little influence on Ecuador–Peru hostilities.

Clear evidence exists that disaster-related activities can and do exacerbate conflict in some cases. The case studies are split amongst disaster-related activities exacerbating conflict, not influencing conflict, and diminishing conflict. The key elements of the disaster-diplomacy hypothesis seem likely to apply to, but need to be tested further for, disaster-related activities exacerbating conflict.

8 Spin-offs

As alluded to when discussing the definitions of 'disaster' and 'diplomacy' (sections 1.1 and 3.1), expansions of disaster-diplomacy work include broadening the understanding and applicability of the terms 'disaster' and 'diplomacy'. These extensions lead to spin-offs from disaster diplomacy, employing similar approaches to slightly different fields of study.

Two spin-offs are described in this chapter. One spin-off, environmental diplomacy (section 8.1), emerges from disaster-related activities covering wider environmental management topics than only disasters. The other spin-off, disaster para-diplomacy (section 8.2), covers non-sovereign jurisdictions dealing with international relations. These spin-offs also indicate how the theory developed for disaster diplomacy in previous chapters can be applied to topics beyond the strict disaster-diplomacy studies.

8.1 Environmental diplomacy

The term 'environmental diplomacy' has numerous applications in science, policy, and practice. One of the wide fields of environmental diplomacy is examining international environmental agreements, especially with respect to success factors and failure factors (for example, Benedick, 1998; Susskind, 1994). Before that, a rich history exists in authors looking at how environmental factors influence international politics (for example, Sprout and Sprout, 1957).

In the context of disaster diplomacy, environmental diplomacy can be taken one step further. It determines how and why international environmental initiatives do and do not lead to cooperation beyond those initiatives. Two examples are explored here: the Antarctic Treaty System (ATS) and environmental management of the Spratly Islands.

The ATS covers the areas of the Earth that are south of 60°S. Three separate agreements comprise ATS:

- The Antarctic Treaty signed in 1959 along with the Protocol on Environmental Protection to the Antarctic Treaty signed in 1991.
- The Convention for the Conservation of Antarctic Seals signed in 1972.

- The Convention on the Conservation of Antarctic Marine Living Resources signed in 1980.

ATS negotiations bring together treaty members in a forum that is deliberately meant to remain uninfluenced by conflicts external to the system. Have positive diplomatic results occurred outside the ATS area due to the ATS negotiations? That would emulate the disaster diplomacy pathway 'Focusing on disaster, not diplomacy' (section 6.1). In the end, focusing on the ATS rather than diplomacy has not yielded palpable or direct results outside of ATS.

For instance, in 1987, North Korea joined ATS as an Acceding State, which means being a non-voting member. Representatives from the country do not always turn up to meetings and no impact on their international relations or diplomacy has been claimed. Overall, the ATS seems to yield little spillover into other diplomatic arenas. To avoid the pitfalls of the 'avoiding diplomacy' and 'distraction' pathways that inhibit disaster diplomacy (section 7.1), it is possible that the ATS works by keeping all ATS discussions about the Antarctic area only and by deliberately not linking it to other topics.

Three factors favour the ATS working within such contexts. First, Antarctica is hard to travel to, in terms of logistics including the expense and time needed. Second, no indigenous intelligent beings exist on the continent. Third, the mineral resources – currently focused on oil and gas – are not currently economically extractable.

If any one of those three characteristics were to change – with the first and the third being the most likely candidates – then the ATS could be threatened. That is, the ATS might be successful because the area that the treaty governs is hard to reach and little of economic value exists there. The lack of propinquity (section 5.2.2) between the ATS area and treaty members perhaps precludes possibilities for environmental diplomacy. The isolation that keeps the ATS separate from other conflicts is the same isolation that permits the ATS work, effectively paralleling the 'Avoiding diplomacy' disaster diplomacy pathway (section 7.1). With changes to that isolation, spillover is possible in terms of environmental conflict, especially over mineral resources, rather than for environmental diplomacy.

Outer space and the deep sea are two other regions with similar characteristics to Antarctica. Possibilities for international cooperation leading to diplomatic results – or perhaps to violent conflict – exist in a similar manner to Antarctica for similar reasons.

Environmental diplomacy analyses need to seek spillover from international negotiations about those regions to determine whether or not cooperation is generated beyond the specific environmental initiative. The alternative conclusion would be to emulate the isolation of ATS, by ensuring that outer space and the deep sea are not connected to other topics. That could permit successful international negotiations and agreements governing those areas.

The second example of environmental diplomacy explored in this chapter is the Spratly Islands in the South China Sea. The Spratlys or parts of them are claimed by six countries: Brunei, China, Malaysia, the Philippines, Taiwan, and Vietnam.

Fortes (2002) suggests making the island group and the surrounding sea a biosphere reserve. It could potentially be jointly managed by the six countries in order to see if tensions in the area would be reduced through managing biodiversity and ecosystems. The idea would be effected by creating a regional body for managing the islands, focusing on cooperation with respect to the area's ecology while providing international support for the parties to succeed in their collaboration. That attempts to play propinquity (section 5.2.2) as an advantage for active environmental diplomacy.

Fortes (2002) cites the precedent of joint efforts between Russia and Japan to conserve natural resources on the disputed Kuril Islands. The Peace Parks initiative (Ali, 2007; Duffy, 2001), which has been proposed for the Kuril Islands as well, is another example.

Peace Parks aim to set up regions of natural-resource management and conservation across international boundaries in order to bring together people from different countries for common park-management purposes. Conflict and non-conflict zones are included, using an active form (section 5.2.7) of environmental diplomacy through propinquity (section 5.2.2), mainly at a state level (section 5.2.6).

Peace Park benefits cited are environmental conservation, including supporting animal migration routes, and livelihood generation such as from tourism and environmental management jobs. Concerns have been raised regarding loss of border control which can favour smuggling goods (including drugs and weapons) and people as well illegal crossings by migrants. Multiple purposes (section 5.2.5) interact for and against Peace Parks. Whether or not genuine conflict resolution results emerge, and whether or not any such results impact wider rapprochement processes, has yet to be determined.

For the Spratly Islands, three main factors work against some or all of the six states cooperating to preserve the ecosystems. These factors are the potential presence of oil and gas reserves, the strategic importance of the shipping lanes, and national pride. Could environmental diplomacy overcome these factors and ensure cooperative resource management and biodiversity/ecosystem preservation?

The first issue is whether or not these two goals are compatible. In particular, could the islands' ecosystems avoid significant damage if the shipping lanes are used and/or if fossil fuels are extracted? Alternatively, the six countries might end up agreeing to share the fossil fuel reserves and shipping lanes at the expense of the ecosystems – the converse of environmental diplomacy.

The second issue is that multilateral cooperation for natural resources frequently does not have a robust basis, even amongst friendly states. Canada and the USA, for example, compete over lumber (Yin and Baek, 2004) and freshwater (Malfatto and Vallet, 2004).

122 *Spin-offs*

As with 'disasters worsening relations' for a disaster diplomacy pathway (section 7.1), environmental diplomacy opportunities can worsen inter-state relations. Expecting six diverse countries with a history of conflicts to join together over fossil fuels, shipping lanes, or ecosystems might be having too high hopes. Foisting those expectations onto those states might doom any environmental diplomacy desires, as per disaster diplomacy's 'Expectations' pathway (section 7.1).

Nonetheless, such an effort could yield successes if similar pathways to disaster diplomacy (section 6.1) were actively pursued. The main ones might be 'informal networks', 'science', and 'symbolism'. Then, subsequent factors must be considered. That brings the discussion closer to disaster diplomacy by considering the disaster of climate change affecting the Spratly Islands.

As described in section 4.14, climate change has the potential for significantly changing many low-lying islands. Even if complete inundation does not occur, then geomorphological or freshwater changes could still undermine existing ecosystems on the Spratlys.

Under the circumstances that some form of a Peace Park or biosphere reserve were created around the Spratlys, leading to cooperation amongst the claimant countries on wider topics, climate change ruining the Peace Park might also undermine the basis for the cooperation. That is, if collaboration is founded on the park but the park loses its viability, then what happens to the collaboration? This possibility is similar to disaster diplomacy's suggestion that diplomacy founded on the basis of disaster-related activities could lead to tenuous diplomacy (section 7.1).

Differences arise between forms of environmental diplomacy and disaster diplomacy. Most straightforwardly, environmental management regimes are generally welcomed or are seen to be positive, irrespective of spin-offs. In contrast, disasters, unless used as weapons (sections 5.2.7 and 7.1), are normally undesired and are seen to be undesirable, even though benefits can be sought from the devastation.

8.2 Para-diplomacy and beyond

Another spin-off of disaster diplomacy is 'disaster para-diplomacy'. Para-diplomacy is international affairs and international diplomacy conducted by non-sovereign jurisdictions, such as cities, provinces, or overseas territories. Examples of non-sovereign jurisdictions are Kuala Lumpur (a Malaysian city), Saskatchewan (a Canadian province), and the Turks and Caicos Islands (a UK overseas territory).

Disaster para-diplomacy explores disaster diplomacy for non-sovereign jurisdictions. That is, sub-national entities interacting directly with foreign governments or international organisations for disaster-related activities. Para-diplomacy frequently occurs for culture, sports, and trade which gives the pre-existing diplomatic basis potentially needed for disaster para-diplomacy to occur.

Kelman et al. (2006) investigated disaster para-diplomacy for case studies of non-sovereign islands of the Commonwealth. Examples are Prince Edward Island (a Canadian province), Tasmania (an Australian state), and Zanzibar (one of the political entities comprising the United Republic of Tanzania).

Legal regimes were examined through the empirical evidence of countries' constitutions, disaster-related legislation, and inter-governmental organisations, such as the UN. Circumstances of disaster-related para-diplomacy are not mentioned in those documents, so legally, it is neither permitted nor denied. Overall, legal mechanisms in Commonwealth non-sovereign islands do not appear to consider the possibility for disaster para-diplomacy.

A few isolated examples were found where it happened. Several disaster-related projects are implemented by international agencies on non-sovereign island territories including Anguilla, the Cook Islands, Montserrat, and Niue. The international agencies deal directly with the island governments. Meanwhile, several Caribbean and Pacific regional agencies provide multilateral fora for disaster-related topics in which these non-sovereign island governments participate. These examples demonstrate the pathways supporting disaster diplomacy of 'informal networks', 'multiple levels/tracks', 'multi-way process', and 'science' (section 6.1).

Despite these few examples, many more possibilities arose where disaster para-diplomacy could have been enacted, but was not. Infrastructure on Tristan da Cunha, a dependency of the UK overseas territory of St Helena, was severely damaged by a storm on 21 May 2001. Despite perceptions that the UK government did provide adequate assistance to Tristan da Cunha, no intimation was raised of going to other governments or to international agencies.

Montserrat, a UK overseas territory in the Caribbean, was badly treated by the UK government in terms of assistance after a volcano starting erupting in 1995 destroying much of the island's infrastructure and leading the majority of the population to leave (for example, Clay, 1999; Pattullo, 2000). Para-diplomacy was a rarely enacted option. The general reaction amongst Montserratians was that the UK should be doing more to assist, rather than considering independence or lobbying others to help.

In 2004, one resident of the Cayman Islands, a UK overseas territory in the Caribbean, invited the USA to intervene to assist with post-hurricane law and order, because the UK was not providing the requested support. That suggestion was derided by other Cayman Islanders. Disaster para-diplomacy was not of interest. Even using the 'symbolism' pathway (section 6.1) to promote disaster para-diplomacy was of little interest.

Why does the disaster para-diplomacy option exist, but is rarely used? Kelman et al. (2006) articulate why island disaster para-diplomacy is not as popular as other para-diplomacy activities, such as culture, sports, and trade.

Non-sovereign islands enjoy a wide variety of forms of autonomy and find advantages from that autonomy (Baldacchino, 2004; Baldacchino and Milne, 2000, 2009). The historical pattern has generally been for island populations to vote against sovereignty in referenda. In contrast, non-island territories have tended to seek independence.

For non-sovereign islands, the governing state (for example, the USA or the UK) offers a crutch or a lifeline, particularly in times of need. Disasters are seen as being times of need. Rightly or wrongly, the expectation – or perhaps just the hope – is that the governing state will step in to assist.

By this governance logic, no need for disaster para-diplomacy (or for disaster risk reduction) exists, because the governing state will assist after a disaster. Even after the governing state demonstrates little interest in providing full post-disaster support – as seen with the UK government's treatment of the Cayman Islands, Montserrat, and Tristan da Cunha – the non-sovereign island rarely shifts towards disaster para-diplomacy to fill in the gap.

The possibilities for a disaster to affect the desire for independence is not a new aspect of disaster diplomacy. The 1970 cyclone in East Pakistan is mentioned in section 1.2. Additionally, not all such instances influence sovereignty politics. As noted, the Cayman Islands, Montserrat, and Tristan da Cunha have not been pushed towards or away from independence despite the initially inadequate disaster-related responses from the UK government.

Rather than waiting passively to see what impact disaster-related activities might have on independence interests, disaster para-diplomacy could be used actively (section 5.2.7) to promote sovereignty or non-sovereignty. Successful para-diplomacy, such as using external assistance for disaster-risk reduction or for disaster response, could garner support for more autonomy or for full sovereignty. Conversely, reliance on external assistance from the governing state could show the need for dependency, garnering support for opposing independence.

Considering disaster-related loss of sovereignty, no recent case studies were found where an independent state lost its sovereignty due to a disaster. Throughout history, invasions, which could be considered to be disasters, have ended many states' sovereignty. East Timor declared independence in 1975. Nine days later, the disaster of the Indonesian invasion ended its sovereignty which was eventually gained back in 2002.

One close analogy is Nauru, an island state in the south Pacific. Due to phosphate resources, Nauru after its independence in 1968 was one of the richest per capita countries in the world. That wealth was squandered through corruption, mismanagement, and incompetence, egged on by outside interests who cared for only their share of the short-term profits. They took advantage of naivety within the Nauruan decision-makers. For instance, Weeramantry (1992) describes how Nauruans were given and listened to shoddy advice from outsiders regarding investments.

With the phosphate depleted and little money remaining, Nauru retained its statehood in name but for a while was effectively an Australian dependency. Australian officials were managing the government and were trying to create a future for the country. This case study could be interpreted as environmental diplomacy (section 8.1) between Australia and Nauru. It could also be interpreted as the disaster of natural-resource devastation or national financial meltdown leading to effective loss of sovereignty.

Kelman and Conrich (2011) develop a framework for analysing island disaster para-diplomacy. The framework is modified here to extend to all para-diplomacy, beyond just islands. The areas to examine to determine the degree to which disaster para-diplomacy is likely to appear are the non-sovereign territory's:

- Autonomy level and sovereignty interests.
- Transportation access.
- Natural resources; for instance, fish, minerals including fossil fuels, natural heritage, and arable land. That links to possibilities for environmental para-diplomacy, following on from environmental diplomacy (section 8.1).
- Population size.
- Ability and motivation to engage in para-diplomacy of all forms.
- Loyalty to the governing state, especially compared to loyalty to other political entities, sovereign or non-sovereign.

Regarding transportation access affecting disaster para-diplomacy, several characteristics need to be evaluated. Mode and expense of transport for reaching the non-sovereign territory, from both the governing state and from the nearest territory, are important. The distances and travel times involved should also be considered.

Proximity and ease of access sometimes increase amity through knowing and respecting one's neighbour. Closeness can also foster mistrust and the desire to be further separated or deliberately different. As with the discussion on propinquity as a category for disaster-diplomacy case studies (section 5.2.2), propinquity for disaster para-diplomacy does not necessarily yield an obvious pattern. The false propinquity pathway (section 7.1) is absent as well, in that the non-sovereign jurisdictions tend to focus on their governing state for disaster-related activities. They do not fully engage with other states, irrespective of propinquity.

The para-diplomacy spin-off from disaster diplomacy has other dimensions. Literature discusses many forms of non-sovereign international diplomacy. Terms used include proto-diplomacy and micro-diplomacy, with the differences amongst them often being too subtle to distinguish operationally (Aldecoa and Keating, 1999; Bartmann, 2006; Duchacheck et al., 1988).

For example, proto-diplomacy is often considered to be para-diplomacy with overtones of desires for more autonomy or independence. That can be too tenuous to prove or disprove.

Northern Cyprus provides a specific example. Is it possible, worthwhile, and useful to distinguish international diplomacy from Northern Cyprus if that jurisdiction is viewed as a de facto state (possibly yielding disaster diplomacy or environmental diplomacy) in comparison to it being viewed as a de facto province of Turkey (possibly yielding disaster para-diplomacy or environmental para-diplomacy)? Does the answer change as Cyprus moves towards possibilities for reunification? Within the context of such questions, Der Derian's (1987) categories of mytho-diplomacy, proto-diplomacy, anti-diplomacy, neo-diplomacy, and techno-diplomacy could be seen to be satirical in trying to manufacture differences amongst difference styles and forms of diplomacy, even where they are operationally virtually identical.

Disaster para-diplomacy on Commonwealth islands (Kelman *et al.*, 2006; Kelman and Conrich, 2011) illustrates further. In the few instances where disaster para-diplomacy occurred, it was mainly out of necessity to deal with disasters, and sometimes to pursue disaster risk reduction, rather than from a desire to engage in international affairs or to promote sovereignty. The situations, at least, could be interpreted in that manner. Some individuals or organisations might have had different purposes than others, even while they all agreed to the same actions. The process of pursuing certain disaster-related activities could also change views of sovereignty for one's non-sovereign islands.

The challenge emerging is similar to that of the disaster-diplomacy 'purpose' discussion in section 5.2.5. Many purposes, admitted and not admitted, known and unknown, occur in decision-making. Deconstructing those to give them specific labels is not always feasible. In the end, the different diplomacy tracks (section 5.2.4) provide a more pragmatic and useable method of categorisation. It applies to diplomacy from both sovereign and non-sovereign jurisdictions.

9 Limitations

9.1 Ethics

Active and passive disaster diplomacy were discussed in section 5.2.7. The ethical considerations were alluded to in terms of actively aiming for disaster-diplomacy outcomes – and hoping that actions taken were correct for achieving the expected outcomes. Several ethical concerns arose that are detailed in this section.

In particular, if a situation emerges where disaster-related activities are seen to have a strong potential to positively impact diplomacy, would methods exist for ensuring that the catalytic effect is made to appear? If such methods exist, should those methods be used, especially in terms of an active choice to apply those methods?

These questions apply to disaster-diplomacy research as well as to the operational decision making. One important question regarding disaster-diplomacy research ethics is whether or not carrying out disaster-diplomacy research could interfere with disaster-diplomacy-related action.

Disaster diplomacy is a highly political topic, namely because disaster-related activities and diplomacy are highly political topics. Interpretations and questions manifest which relate to the morality of humanitarian aid and other disaster-related actions. For example, rather than starting with 'do' or 'can', disaster diplomacy's defining question (section 3.1) could start with 'should' giving 'Should disaster-related activities induce cooperation amongst enemies?' If so, then how and why?

The question could be further refined to become 'Under which circumstances should disaster-related activities induce cooperation amongst enemies?' Yet that is a leap forward from the 'should' question. Answering 'yes' to the yes/no 'should' question is not straightforward.

Those who answer 'no' explain that extensive effort occurs to divorce disasters from politics, including scientific and technical cooperation for all disaster-related activities. New mechanisms for relating disasters and politics are not needed. Instead, encouraging further separation would be preferable.

Those who answer 'yes' explain that disasters are inherently political through the process of creating and maintaining vulnerability (for example,

Lewis, 1999; Wisner *et al.*, 2004). The more positive outcomes from disaster-related activities which can be fostered, the better. Such outcomes should be actively pursued.

Exploring principles, policies, and practices with respect to this discussion would be a valuable extension to disaster-diplomacy ethics. That includes during a disaster in order to determine how decision-making processes occur at moments of crisis. Carrying out this research would be best pursued by interviewing policy-makers and decision-makers, including politicians and senior civil servants, potentially during a crisis.

That could draw attention away from dealing with the crisis, possibly causing harm. That would nonetheless help to understand why decisions are made while those decisions are being made. Informing policy-makers and decision-makers of possible disaster-diplomacy consequences and explaining where previous instances of possible disaster diplomacy have failed or succeeded could cause them to alter their decision. Such an outcome is not necessarily detrimental, yet it interferes with the decision-making process.

Disaster diplomacy could also be a distraction during a crisis, as illustrated by the Ethiopia–Eritrea case study (section 4.8). Could the spectre of disaster diplomacy have influenced the decisions and words of Ethiopia and Eritrea? Disaster diplomacy presents an opportunity for diplomacy – yet that could be a threat if rapprochement were opposed.

Suggesting that a humanitarian imperative necessitates certain political actions could also result in a country refusing to accept that humanitarian imperative. Reasons could be loss of face, expected loss of face, not willing to be dependent on an enemy, or wishing to cause further problems for a foe (see also section 7.1).

Conversely, if the Eritrean humanitarian corridor had been implemented, could it have provided a basis for a long-term solution to the conflict? It might have illustrated the advantages of cooperation, rather than keeping people alive temporarily on humanitarian aid while letting the conflict drag on? If non-Eritrean routes were indeed more efficient for delivering aid, as claimed by Ethiopia, could the long-term diplomatic gains of using Eritrean routes be weighed against the short-term humanitarian losses? How would politicians and aid agencies have reacted if those questions were posed publicly and explicitly while the decisions were being made?

The insights would have been fascinating, yielding innovative results and potentially startling conclusions about political decision-making, managing humanitarian crises, and disaster diplomacy. Would that research justify the potential interference with the disaster-related activities, even if the outcomes were applied for more gain in future situations?

These ethical dilemmas apply to publishing disaster-diplomacy material, such as this book and media reports. As seen in the Greece–Turkey (section 4.7) and Iran–USA (section 4.2) case studies, the media often publicise the potential for disaster diplomacy. They knowingly or unwittingly push the topic onto the agenda, forcing public responses. By drawing attention to disaster diplomacy, or lack thereof, continuing in ignorance is no longer an option. Action, positive or negative, is needed.

Glantz (2000) suggests this situation regarding Cuba–USA disaster diplomacy by noting that disaster-related scientific visits between those two countries 'are too numerous to mention and their success may be related to the lack of explicit attention drawn to them' (p. 243). From a research perspective, should a reviewer of Glantz's work accept his statement of the extensive scientific exchange exactly as he presents it? Alternatively, should specific examples be demanded and published in order to produce an appropriately rigorous and referenced scientific publication, even if the Cuba–USA scientific exchange could be undermined?

In the Greece–Turkey case study, Ker-Lindsay (2000, 2007) highlights that diplomatic negotiations between those two 'enemies' had been ongoing before the earthquakes. The disasters thrust the process into the spotlight, resulting in demands from the public for swift diplomatic action. The diplomats were not ready to meet those demands. The rapprochement might have been hampered.

The roller-coaster nature of Greek–Turkish relations in subsequent years supports Ker-Lindsay's (2000, 2007) analysis (see also Mavrogenis, 2009). Yet earthquakes in Turkey, such as on 1 May 2003 which killed at least 100 people, brought offers of help from Greece. Those offers were readily accepted with no indication or expectation stated of diplomatic outcomes. Greek–Turkish mutual humanitarian aid had become accepted as normal, irrespective of political shenanigans (Ganapti et al., 2010).

In the absence of the 1999 earthquakes and the attention given to the Greek–Turkish thaw, would Greek–Turkish rapprochement have been much smoother? Would a major earthquake in Cyprus in 2003 have spurred on the failed 2004 reunification efforts or damaged them further through expectations of disaster diplomacy?

Would publishing and publicising any disaster-diplomacy potential cause the parties involved to react even more adversely to each other and to close possible diplomatic doors in order to avoid the appearance of any reconciliation due to a humanitarian imperative? This attitude is a strong possibility for:

- The Ethiopia–Eritrea enmity continuing despite the droughts.
- The back-and-forth negotiations between the USA and Cuba regarding American aid to Cuba after Hurricane Michelle in November 2001.
- Iran and the USA taking great care to avoid linking humanitarian aid to long-term political changes following the 2003 Bam earthquake.

The decisions taken in all these instances included an element of responding to calls for disaster diplomacy by various parties involved or observing externally.

That leads to another angle of the ethical limitations of dealing with disaster diplomacy. Should disaster-diplomacy commentators, including researchers, take responsibility for the operational outcomes of their publications and calls for action (or inaction)?

In some instances related to humanitarian relief, suggestions of disaster diplomacy have led to strong, adverse reactions about linking disaster-related activities and diplomatic activities. In researching Warnaar (2005), that author was thrown out of one practitioner's office when the practitioner refused to consider the possibilities for linking disaster relief with politics. The general argument is that disaster relief should not be linked to politics. Helping other people survive in a post-disaster context is an obligation. It adheres to principles of fundamental human rights.

A subtext often articulated is the suggestion mentioned above that aid agencies spend time and effort trying to separate politics and disaster relief. Another mechanism for integrating them is unwelcome and could be damaging.

This view from operational aid workers is encouraging because it suggests a strong movement to assist people, irrespective of politics. On the other hand, this attitude could suggest a naïve unwillingness to admit political realities, especially the high-level behind-the-scenes manipulation and political games which some disaster-diplomacy case studies have identified.

As noted above, one view is that, irrespective of how much disaster workers would wish otherwise, disasters and aid are fundamentally political. Hence, opportunities should be grasped to achieve constructive political gains rather than continually becoming mired in negative political consequences.

The extension to this argument is that openly using humanitarian aid for political leverage is not only ethical but is also desirable. Disaster-diplomacy research and practice should entail interviewing decision-makers while aid decisions are being made.

Disaster diplomacy and methods for manipulating aid for positive political gain could be explained to one group of decision-makers while another group would receive no such instruction. The consequences could then be monitored and compared. As well, the level of insight displayed by decision-makers into linking humanitarian crises with long-term political reform could be assessed.

Such an applied-research project would advance disaster-diplomacy work, but it could backfire. A long history exists demonstrating that not all leaders see their duty as being to help others, including their own people, after a disaster. If a leader decided to take advantage of their enemy's weakness following a disaster, and specifically identified the disaster diplomacy researcher or practitioner as having given them the background and evidence necessary to make this decision, what responsibility does the disaster-diplomacy researcher or practitioner assume for having meddled in politics? If the proposals for using natural hazards as weapons of war (sections 5.2.7 and 7.1) are played out on the battlefield and the commander's decision is traced back to this book, rather than to the vast military literature on this topic, does this author bear any responsibility?

Research into these questions as part of the ethical investigation of disaster diplomacy could bring researchers and decision- and policy-makers closer together. Such research in conjunction with practitioners could further provide an improved understanding of how political decisions pertaining to disaster-related activities are made and how to improve those decisions.

9.2 Confounding factors

The evidence presented throughout this book is that disaster diplomacy tends to display few and limited successes, that predicting the outcomes of case studies is challenging other than the general conclusion of failure, and that a general theoretical model might not be developable or applicable. The overall reason is the number and diversity of various influences on disaster-related activities, on diplomacy, and on their connections.

Disaster-related activities are influenced by numerous, complex factors over all space and time scales, leading to numerous confounding factors in trying to focus on disaster diplomacy dimensions. Those influences link directly to wider societal concerns such as human rights, environmental management, development, and sustainability.

In particular, the history of disaster-related research shows that, even when an environmental event is involved such as a hurricane or a volcanic eruption, the disaster does not result from that environmental event. Instead, the disaster results from human actions, decisions, behaviour, and values that created and perpetuated the vulnerability to that environmental event (see more discussion in section 3.1).

Examining the example of earthquakes, most earthquakes happen rapidly, with shaking lasting seconds or, at maximum, minutes. The majority of casualties is caused by infrastructure collapsing (sometimes in earthquake-induced landslides) or objects such as furniture moving, rather than by the earthquake itself. That is the reason for differentiating between earthquakes, the shaking, and earthquake disasters. The latter refers to casualties, damage, and disruption.

More specifically, extensive knowledge exists for designing earthquake-resistant structures, engineered and non-engineered, and for dealing with non-structural aspects of earthquake damage minimisation (for example, Coburn *et al.*, 1995; Dixit, 2003; GHI, 2006). That knowledge has frequently been applied successfully in practice. These lessons and the knowledge are not always transferred to other locations or implemented fully, mainly due to political reasons (for example, Bosher, 2008; Lewis, 2003).

These political reasons require years to manifest. They relate to building codes and planning being or not being developed, implemented, monitored, and enforced. Individual, institutional, and cultural choices need to be made and engrained in order to lead to corruption, incompetence, and apathy.

The earthquake and the shaking are natural and are a normal environmental phenomenon. Understanding why earthquake disasters occur requires deeper explorations into the creation of the political disaster that permits earthquakes to cause casualties, damage, and disruption. In summary, disasters should be labelled as human disasters, rather than as natural disasters (O'Keefe *et al.*, 1976), although exceptions exist.

This understanding of disaster extends to society's actions over the long term and to the multitudinous influences on society's actions, particularly how one segment of society tends to affect the vulnerability of other segments, with or without their consent. Consequently, the challenge of considering all factors becomes apparent. Day-to-day life and decade-to-decade societal decisions are linked to vulnerability and to disaster. These factors can and do confound analyses seeking to establish patterns in influences on and by disaster-related activities. That makes explanatory and predictive models for disaster diplomacy difficult to develop.

Diplomacy yields a similar array and diversity of confounding factors (for example, Bayne and Woodcock, 2007; Chasek, 2001; Diamond and McDonald, 1993; Lecours, 2002). With diplomats and diplomacy dealing with topics from visa agreements to tax treaties, and from resource sharing to espionage, establishing clear links amongst topics can be challenging.

As an example, in March 2009, a disaster-risk reduction conference in Iran did not have a talk from an invited American keynote speaker because the Iranian government refused to grant a visa to the speaker. An opportunity for knowledge exchange to improve disaster risk reduction was lost. In parallel, many academic conferences and scientific exchanges have moved outside the USA due to significant challenges in securing visitor visas that emerged after the 11 September 2001 terrorist attacks in the USA.

International negotiations and treaties on water resources (an aspect of diplomacy) can affect floods and droughts (which could become disasters). Some countries, for instance India (Jayaraman, 2005; section 4.11), have previously been reluctant to share all their real-time seismic data which affects tsunami warning possibilities.

In examining disaster diplomacy purposes (section 5.2.5) and pathways (Chapters 6 and 7), the disaster-related and diplomatic complexities are revealed. Most link to active disaster diplomacy (section 5.2.7) in terms of specific decisions being made to pursue or not to pursue elements within disaster diplomacy. The rationales, or often lack thereof, that enter into those diplomatic decisions lead to a multiplicity of connections amongst all facets of diplomatic activities and diplomatic decision-making processes.

With such extensive connections amongst disaster-related activities along with such extensive connections amongst diplomatic activities, working out the interactions between disaster-related activities and diplomatic activities becomes a huge challenge. Each case study displays its own sets of contexts and numerous factors influencing those contexts. In trying to seek patterns amongst disaster-diplomacy case studies, along with explanatory and predictive models, these interactions become confounding factors in the analyses.

Data quality further confounds analysis. Section 5.1 discusses how much of the quantitative data available might not be good enough for the analyses that are run, making the conclusions suspect. Efforts are ongoing to improve the data within certain sectors (Gleditsch *et al.*, 2002; Harbom and Wallensteen, 2009). Those efforts should be commended and encouraged. They cover only a small portion of the data that would be needed for comprehensive disaster-diplomacy analyses through quantitative means.

Qualitative data are also subject to limitations. Bird (2009) details challenges and limitations of survey-related data within disaster research. Disaster-diplomacy studies have demonstrated specific examples in practice.

In the research reported by Warnaar (2005), some interviewees refused to discuss the topic of disaster diplomacy. To some degree, that is an important result in itself, regarding interest in and the impact of disaster diplomacy. Without the individuals' explanations for why they would not wish to discuss the topic, explanations and analyses are limited.

In the realm of diplomacy, many decisions are made secretly. Even open decisions can be made without explicit explanation, justification, or reason. Interviews provide people's perspectives. Care must be taken in assuming that the information given in interviews is complete, is honest, or is correct.

People might not always know, be able to explain, or be willing to think through why they acted in a certain manner. Decisions made or words said in the heat of the moment might result as much from lack of sleep or uncaffeinated drinks as from careful thought drawing on a decision-maker's years of experience. The outcome is that, sometimes, the most important data are the data that cannot be truly acquired or ever be known.

These challenges are well documented in scientific literature covering field methods, including interviews and observations. They apply to disaster diplomacy. Disaster diplomacy research must proceed with caution while explicitly delimiting the usefulness and usability of any results that are obtained. The confounding factors in any analysis leading to limits of that analysis should always be indicated.

9.3 Bias

In investigating and interpreting disaster diplomacy, care is also needed to try to avoid being influenced by pre-conceived objectives. Few circumstances exist, particularly from researchers and staff working on disaster-related activities, where people would admit to desiring more disasters, more conflict, or less peace. Even at political and diplomatic levels, such calls are rarely heard openly, even if discussions behind closed doors actively promote such policies.

As per Obama (2009), even those advocating war usually do not claim that it is a preferred option. Instead, they suggest that they support violent conflict in specific circumstances because it is a pathway to peace and to their view of security. Also, they state, war is less evil than avoiding violent conflict at all costs.

Thus, a strong bias is usually present as the starting point in disaster-diplomacy studies that fewer disasters and increased diplomacy would be preferred. This starting point is akin to the descriptions in section 9.1 of the different views regarding the question 'Should disaster-related activities induce cooperation amongst enemies?'

In researching disaster diplomacy, it is important to consider the evidence available rather than seeking evidence to conform to pre-conceived conclusions. Those already established conclusions might be the desire for neutral disaster relief, the hope that disaster diplomacy will be successful, or that anyone can be convinced of promoting peace.

In promoting the science with policy- and decision-makers, a possibility exists for the success or failure of disaster diplomacy to become a self-fulfilling prophecy depending on what outcome was sought and pursued. That is, if an active choice is made to pursue and support disaster diplomacy, then elements of success are likely to be witnessed due to that effort.

Post-tsunami Aceh is an example (section 4.10). The tsunami disaster created the space in which peace was a possibility. The parties to the conflict and external mediators supported the peace process within the context of tsunami reconstruction. That led to a reasonable level of success in both processes, although no claim is made that the Aceh situation is perfect with respect to tsunami reconstruction, post-conflict reconstruction, or their interaction.

In contrast, when an active choice is made to inhibit disaster diplomacy, then that choice is likely to spell the end of any disaster-diplomacy possibilities. Iran rejecting the high-level delegation from the USA after the 2003 earthquake is an example (section 4.2). Disaster diplomacy was actively inhibited and hence failed.

Meanwhile, passive disaster diplomacy is likely to yield the self-fulfilling prophecy of few obvious successes, because little effort is put into the disaster diplomacy. Considering Southern Africa from 1991 to 1993 (section 4.4), preventing the drought disaster despite the drought emergency was feasible due to the recent political changes. That illustrated success in the political cooperation. Little disaster diplomacy appeared, because dealing with the disaster had limited influence on the ongoing political changes and peace processes. Similarly, vaccine diplomacy (section 4.16) has had successes regarding dealing with disease, but few tangible political outcomes.

Often, short-term results appear regarding disaster diplomacy which biases towards assumptions of success. Following the 2001 Gujarat earthquake (section 4.9), a six-month honeymoon was enjoyed in India–Pakistan relations. Both sides appeared to genuinely seek reconciliation that had emerged from the earthquake. Disaster diplomacy seemed to have happened.

The process soon collapsed, setting back collaboration between the two countries. That lost ground was promptly recovered, but most advances were made in the absence of disaster-related activities. Potentially due to experience from 2001, the 2005 Kashmir earthquake neither derailed nor extensively boosted the slow, careful manoeuvring enacted by both India and Pakistan in seeking increased connections and collaborations.

10 Principal lessons for application

While being cognisant of the limitations of disaster diplomacy, it is also important not to be entirely hampered by those limitations. The research can and should still be consolidated to obtain lessons that could be used by practitioners, in both disaster-related and diplomatic activities. That might help avoid the often standard, yet incorrect, assumption, that disaster diplomacy can and should work. That is, the undue optimism that often pervades disaster diplomacy needs to be countered.

The most important bias to consider is the assumption that fewer disasters and increased diplomacy would always be preferred by all parties involved (section 9.3). The lessons here adopt that bias as a starting point. They aim to inform those seeking fewer disasters and increased diplomacy how disaster diplomacy does and does not work; that is, how disaster-diplomacy research can and cannot be used.

The lessons examined here are:

- Be ready for assistance offers from enemies.
- All diplomacy tracks can be useful.
- Disaster diplomacy operates at many levels.
- Lessons should be implemented, not forgotten.

Common threads run throughout each of the four lessons: multiple tracks and diversity that interconnect the complicated disaster-related and diplomacy processes. These threads are the theme of Kelman (2010). Even when boosting disaster diplomacy is not at the top of political agendas, and even when the most prominent aspect of disaster diplomacy is that it did not work, the tangled interconnectedness throughout the multiple tracks leads to continuing opportunities for potential disaster-diplomacy successes. The separate lessons presented here indicate that, despite the apparent failure of disaster diplomacy in most instances, significant hope remains that applying the lessons could produce more encouraging results – if policy decision-makers choose to pursue successful disaster diplomacy.

10.1 Be ready for assistance offers from enemies

One of the most notable aspects of the Hurricane Katrina disaster-diplomacy case study (section 4.12) was the US government's actions regarding aid (Kelman, 2007). The US government displayed inconsistent reactions to international aid that was being offered, stumbling on whether to accept or reject the proffered assistance.

An absence of planning was evident along with a lack of pre-disaster consideration that the USA might need external disaster-related aid and might need to deal with offers, needed or not. The US House of Representatives (2006) described how 'Ours was a response that could not adequately accept civilian and international generosity' (p. 2). GAO (2006) criticised the US government's inability to deal with external offers.

The lesson is to include offers of assistance, from enemies and friends, in any disaster-related planning.

When the UK's Civil Contingencies Bill (House of Commons, 2004) was drafted and opened for public consultation in 2003, ACNDR (2003) suggested in its submission that the need to call for international assistance should be acknowledged in the bill and prepared for. That did not happen. The final legislation did not address the potential for the UK to need, request, or receive international post-disaster assistance.

Equivalent legislation in New Zealand (National Civil Defence Emergency Management Plan Order, 2005, p. 57) specifically highlights the possibility that New Zealand might call for or request international assistance after a disaster:

> International assistance for New Zealand
> 79 Introduction
>
> (1) An emergency in New Zealand may generate offers of assistance from overseas governments and non-governmental organisations, or necessitate requests from New Zealand for external help.
> (2) The Government will address requests for, and offers of, overseas assistance through the government crisis management arrangements of DESC.
>
> 80 Requests for international assistance
>
> (1) The Government may request international assistance in a civil defence emergency.
> (2) The National Controller or the Director will seek approval for the deployment of international assistance.
> (3) MCDEM may require international support to co-ordinate the entry and deployment of international assistance.

It is straightforward for other countries to include such text in national emergency plans.

Exercises dealing with national-scale emergencies should include the possibility that unfriendly parties might offer disaster-related assistance. Consequently, guidelines would be useful regarding how to address these offers. For instance, offers of aid from groups designated as terrorist organisations might be ignored or summarily rejected. Conversely, all official offers from sovereign states or other official governments should be acknowledged, whether or not that aid is accepted. One proviso might be that a government should have been democratically elected to receive official acknowledgement.

The possibility of disaster-related assistance from enemies does not apply to only post-disaster aid. Pre-disaster actions are disaster diplomacy case studies, such as Glantz (2000) examining Cuba–USA information sharing on hurricane modelling (section 4.6).

Where an entity, such as a national government or a local community organisation, is working on disaster topics, decisions might be needed on the level or type of assistance to accept from others. That might be exchanging material over email or joint in-person training exercises. Many reasons might exist for disaster-related collaboration offers being made or being accepted, from espionage to desiring to establish a baseline for reducing enmity. Care is needed in responding to any offers recognising the disaster-related and diplomacy-related advantages and disadvantages of doing so.

The lesson is still to be ready for assistance offers from enemies. Such offers could occur at any time for any aspect of disaster-related activities. A specific disaster is not necessary, although that is frequently the impetus.

The offer might be spontaneous, such as Venezuela to the USA after Hurricane Katrina (section 4.12). The offer might be part of an ongoing process such as Cuba–USA regarding mutual pre-hurricane and post-hurricane aid (section 4.6) or the Middle East Regional Cooperation Program (section 3.1). The offer might have the potential of being inadvertent if, for example, disaster-related topics are raised as part of a wider package of negotiations.

Planning for these possibilities often assists further by identifying gaps in disaster-related activities, through answering the question 'Why might disaster-related assistance be needed from enemies (or friends)?' Irrespective of mirror disaster diplomacy (section 6.3), answering that question would be a useful impetus towards enacting disaster risk reduction. Offers of assistance from enemies, regarding a specific disaster or not, could also come from any diplomacy track. That is the subject of the next lesson.

10.2 All diplomacy tracks can be useful

Diplomacy is not just about diplomats and political manoeuvring, but can also emerge along many tracks (section 5.2.4). Neglecting or emphasising certain tracks could open the possibility for being taken by surprise by a neglected or unemphasised track.

Greece–Turkey after the 1999 earthquakes (section 4.7) demonstrated how the people and the media dragged along some of the decision-makers to some extent. That exposed an ongoing but low-key rapprochement process.

Relatively few diplomats and politicians tend to be scientists, so the power of scientific and technical collaboration to forge links can be overlooked. The Cuba–USA case study (section 4.6) demonstrates how many of these links were occurring for climate-related disasters when Cuba was led by Fidel Castro. The research collaboration also spilled over into developing science-based policies for disaster-related education and training as well as for operational forecasting.

An example is the 'climate affairs' template (Glantz, 2003) which is a book outlining a structure for training workshops and university curricula to bring together anyone with an interest in climate. The topics covered are climate science, climate impacts, climate policy and law, climate politics, climate economics, and climate ethics and equity. This approach galvanises attention and forces enquiries, bringing together meteorologists and water lawyers along with farmers in arid regions and international development philosophers. Their common interest in climate is used to connect them.

The 'affairs' template was used by American scientist Michael H. Glantz (author of Glantz, 2003) and his Chinese colleague Qian Ye to set up the International Center for Desert Affairs in Urumqi, Xinjian, China. The center was established and continues to operate despite ongoing, high-level disputes between the USA and China over topics from dealing with North Korea to trade barriers. Furthermore, one of the main research fields for the centre is 'Diversity of Culture', despite the high political tensions over multiculturalism in Urumqi (Dillon, 1997).

The Desert Affairs centre and programme have not had any discernible impact on reducing these tensions or on bringing together Beijing and Washington, DC, but that is not the mandate. The centre and programme forge scientific and technical links between the two countries for dealing with disaster- and environment-related changes. That represents one of diplomacy's tracks.

Back to Cuba–USA collaboration, the 'affairs' template generated collaboration between Glantz and several Latin American scientists, including a Cuban meteorologist. One outcome was the development of El Niño Affairs in Spanish for Latin America, covering climate, the environment, and society. That programme involves collaboration between Peruvian and Ecuadorian scientists, which started at a time with simmering hostilities between those two countries.

Disaster-diplomacy successes at the scientific and technical levels can occur because diplomats and politicians are not involved in, or are unaware of, this work (Glantz, 2000). A danger exists that highlighting initiatives or bringing this diplomacy track to the attention of decision makers could scuttle it. Section 9.2 described how visa applications for disaster researchers to visit colleagues or to attend international meetings are sometimes declined.

Consequently, no diplomacy track can be relied on. That is particularly true for the official diplomatic channels in that some of the examples demonstrate how resolving disaster-related transboundary concerns might not be dominated by governments. Even if the diplomacy track is open, it might not be successful or it has the potential for backfiring. The lesson here is to rely on neither a single track nor a single set of tracks to achieve diplomatic aims.

Instead, all diplomacy tracks can be useful. Extensive diplomatic work can be achieved beyond actions from diplomats and politicians.

How does this lesson apply in practice? Section 5.2.2 speculated regarding disaster diplomacy from floods or flood-warning systems along the Congo River in the Kinshasa–Brazzaville area. Both are capital cities, yet both national governments have suffered from instability. Both countries were tied at #162 out of 180 on Transparency International's Corruption Perceptions Index 2009, indicating a high likelihood of significant corruption.

Local or external groups attempting to set up a joint flood risk-reduction programme between the two sides of the river might choose to work with local partners in both cities, such as NGOs or academic organisations. That would seek success that might not be feasible in working with the national authorities. This situation illustrates that, even when the governmental diplomatic track appears to be best avoided, significant possibilities still exist to pursue disaster diplomacy, if that is desired.

10.3 Disaster diplomacy operates at many levels

Most of the case studies and theoretical analyses of disaster diplomacy describe the general failure of disaster diplomacy. In particular, seeking new, legitimate, long-lasting diplomacy from disaster-related activities addressing long-term issues (section 3.3) might be unfairly ambitious. That rigid, comprehensive approach that aims for 'real' or 'pure' disaster diplomacy might be setting up the analysis to yield failure, thereby doing a disservice to the achievements of disaster diplomacy.

Multiple tracks of diplomacy and multiple tracks of disaster-related activities yield multiple tracks of disaster diplomacy. Abject failure or attempts to stymie disaster diplomacy at one level might obscure shifts forward in, or enthusiastic support for, disaster diplomacy at other levels. In extracting lessons from disaster diplomacy, being biased by short-term results (or short-term lack of results), or being biased by results from one track, could be overly focused on one small component of the overall disaster diplomacy picture.

The Iran–USA case study (section 4.2) provides an example. After the Bam earthquake in 2003, the American attempts to pursue higher diplomatic channels based on earthquake aid failed, effectively ending the disaster-diplomacy possibilities. In the meantime, the American rescuers were forming strong bonds with many Iranians.

That leads to low-level exchanges and continuing links between the two countries for disaster-related activities. Iran–USA relations have worsened in recent years and the high-level earthquake diplomacy is forgotten. In parallel, other forms of Iran–USA cooperation could be setting a baseline at the individual level for future thaws between the two countries, once the politicians permit it.

Similar approaches were implied in the previous recommendation regarding the scientific and technical collaboration that generally occurs on a one-on-one basis, such as with the Cuba–USA case study. While the leaders in Havana and Washington, DC play their political games, people at lower levels, even if government employees, have a professional and personal passion to be collaborating on disaster-related activities. That is frequently not with any designs for large-scale diplomatic revolutions. Instead, it is about doing one's job as a scientist. Sometimes that has an added push from personal friendships or enjoying professional encounters with other like-minded scientists.

Greece–Turkey (section 4.7) evidences the power of different levels of success. Ker-Lindsay (2000, 2007) indicates that thrusting the diplomacy into the spotlight threatened to derail the process. At the diplomatic and political levels, questions were being asked regarding the rapprochement. Those supporting it were targets for those questioning the process.

In spite of that situation, the anti-disaster diplomacy machinations barely enter into Ganapti *et al.*'s (2010) analysis regarding non-governmental disaster-related cooperation. While one track was showing cracks, another track was bolstering the links that the earthquakes had boosted, solidified, or formed.

A diversity of disaster-diplomacy successes and failures within the same case study should not be undervalued. Individual connections, perhaps made in the aftermath of a disaster, can make a difference in the long run if those individuals end up in more powerful positions. This discussion is part of the discourse on public diplomacy.

One of the limitations of many of the case studies explored here is the short-term nature – usually under a decade. While the influence of disaster-related activities seems to fade over the long-term in the shadow of non-disaster factors (section 3.2), sometimes disaster-related tendrils have long reaches, which could be applied rather than dismissed or forgotten.

10.4 Lessons should be implemented, not forgotten

One of the major challenges of listing lessons is how to ensure that they are heeded. 'Useable science' (Glantz, 1997) does not necessarily mean 'used science'; 'applied science' does not always mean that the science is applied or is applied appropriately. Science might be provided in a form yielding lessons to be applied, but that does not inevitably lead to the science being used. Achieving the goal of lessons implemented, rather than just lessons learned, might not always be feasible (Glantz, 2008).

Disaster diplomacy can be particularly irritating from the perspective of ensuring that lessons are genuinely learned and implemented, rather than learned and then forgotten. The reason is that disaster diplomacy is not always the goal of those involved.

The Cuba–USA (section 4.6) and Eritrea–Ethiopia (section 4.8) case studies explained how both sides put in effort to ensure that disaster diplomacy did not occur. Section 7.1 explains the pathways inhibiting disaster diplomacy and how those pathways can be actively pursued as part of active disaster diplomacy (section 5.2.7), but actively inhibiting disaster diplomacy.

As noted in section 9.3, the goal of those making decisions might not even be either disaster-risk reduction or diplomacy. Without some form of impetus towards avoiding disaster and towards creating diplomacy, lessons on supporting disaster diplomacy are not relevant. In particular, the lesson regarding being ready for assistance offers from enemies (section 10.1) intimates that even if disaster diplomacy is not of interest, advantages exist in accepting its possibility and in planning for that eventuality. That could encourage some of those involved to plan specifically to oppose disaster-diplomacy possibilities.

Although this analysis about implementing lessons seems bleak, a method exists for promoting lessons for implementation. Since diplomacy and disaster-related activities occur along multiple tracks, it would be unlikely that all tracks simultaneously have similar goals. It is likely that some tracks have interests in disaster diplomacy while others ignore it or actively seek to avoid it. For those seeking to learn and implement disaster-diplomacy lessons, the supportive tracks should be sought.

By carefully targeting efforts towards those who are predisposed towards supporting disaster diplomacy, rather than necessarily directly opposing factions or tracks inhibiting disaster diplomacy, opportunities can be created for implementing the lessons. The overall, and significant, caution is that the lessons emerge from only a small set of analyses covering only a small set of case studies.

While it is appropriate to act on the basis of what is known, all decisions must recognise that new evidence has the possibility of changing the analyses and the conclusions. New lessons could be rendered. At the very least, actions on disaster diplomacy should continually be monitored and evaluated to feed new evidence into the decision-making process while striving to fill in the gaps indicated by the future disaster-diplomacy work.

11 Filling in the gaps

Yim *et al.* (2009: 291) explain that 'disaster diplomacy lacks a formal definition of principles, metrics of success, a strategy for integration into formal diplomatic efforts, and a dedicated training programme for humanitarian agents planning to engage in this form of diplomacy'. Some of the research aspects of these gaps are partially addressed in this book, but Yim *et al.*'s (2009) fundamental critiques remain legitimate and the gaps they identify still need to be filled fully.

For instance, the formal definition of principles is covered to some extent in Chapter 3 while integration into formal diplomatic efforts is discussed in section 9.1. None of this work is complete. Yim *et al.*'s (2009) desire for more formal integration of the research and practice are apposite.

11.1 Can the limitations be overcome?

The limitations in the evidence and analyses of disaster diplomacy have been described throughout this book. Those limitations are straightforward to overcome. Further case studies can be investigated to produce more evidence to be input into the analysis. That covers the qualitative typologies (section 5.2) as well as the quantitative analyses (section 5.1).

The quantitative analyses, in particular, have many possibilities for overcoming the limitations. Namely, acknowledging the weaknesses of the data being used, trying to fill in gaps where feasible, and triangulating with multiple methods where gaps cannot be filled.

Extensive literature exists on different forms of, and meta-analyses about, conflict indices, democracy indices, peace indices, disaster indices, and vulnerability indices, amongst other attempts to quantify. Deeper analysis would cover how using different indices produces similar or different results along with sensitivity analyses to see how errors in the data affect the outcomes. Above all, when quantitative correlations are calculated, describing mechanisms for those correlations existing, along with descriptions on why those correlations could be spurious, are important for quantitative analyses.

Another limitation pervades disaster-diplomacy work: whether or not the concept itself is limited operationally. Section 9.1 asked 'Should disaster-related activities induce cooperation amongst enemies?' which led to explanations of answers covering 'yes' and 'no'. Those answers were from a conceptual standpoint. Operational approaches might argue that a conceptual limitation of disaster diplomacy is that it does not significantly assist operational work.

An imposed and artificial separation between disaster and diplomacy could perhaps be used as a powerful force for addressing both disaster-related activities and diplomacy in separate fora. Established and respected modes for humanitarian aid and disaster-risk reduction could be promoted.

A good example is the Seven Fundamental Principles bonding the National Red Cross and Red Crescent Societies, The International Committee of the Red Cross, and the International Federation of the Red Cross and Red Crescent Societies. These seven principles are Humanity, Impartiality, Neutrality, Independence, Voluntary service, Unity, and Universality. The impartiality and neutrality principles support the approach of not connecting disaster-related activities with diplomatic or political demands. Operationally, it might easier to achieve inroads into both topics if they are deliberately and actively separated.

For disaster-related activities, the relevant ideal is that disaster-related work is for all of humanity without prejudice. That implies impartiality and being neutral in any ongoing political conflicts, because the focus needs to be on robust disaster-risk reduction or humanitarian assistance. For diplomacy, no claim of impartiality or neutrality is made, although diplomacy sometimes includes work to create an impartial and neutral space where conflicting parties can negotiate non-violently.

Others have questioned the relevance, practicality, and appropriateness of these principles under all circumstances (for example, Cuny, 1983; Norris, 2007; Rieff, 2003; Terry, 2002). The accusation is not always of naivety, but is about being realistic.

Those who critique the separation of disaster and diplomacy for all situations accept the tenet that all disasters are inherently political, hence disaster-related activities are inherently political processes. The argument is that disaster-related activities cannot be successful or robust without accepting the political causes and solving those political problems. Since disasters and diplomacy are inherently interconnected (section 9.2), perhaps it would be best to admit that reality. Then, that reality could be used to seek positive outcomes from the interconnectedness.

The dangers of that approach are expressed by the pathways that inhibit disaster diplomacy (section 7.1). The lessons from Chapter 10 indicate that many pathways operate simultaneously involving multiple parties. That can be advantageous in focusing on pathways and parties that support disaster diplomacy (sections 6.1 and 10.4).

Politics, by definition, both interferes with and promotes disaster-related activities. It can be constructive to work with the opportunities that promote disaster-related activities and disaster diplomacy rather than confronting the contrarians. That approach could work as long as the contrarians do not become an overwhelming force.

Reconciliation between the opposing views might not be necessary. To overcome the operational limitation of disaster diplomacy, the lesson is to pursue multiple tracks and levels simultaneously (sections 10.2 and 10.3).

Some parties divorce disaster and diplomacy. Some passively accept the connections. Some actively seek to create, pursue, and exploit connections, for multifarious purposes. No rule dictates that a singular view of disaster diplomacy is essential, conceptually or operationally.

Instead, different parties have different interests and needs. It is feasible for opposing approaches to exist side-by-side. As such, disaster diplomacy's research and operational limitations become strengths by permitting some to separate disaster from diplomacy and others to join the two.

11.2 Why further study disaster diplomacy?

Given that disaster diplomacy's research and operational limitations can be overcome, at least to some degree, optimism exists that the lessons from Chapter 10 can and should be applied for future disaster-diplomacy work. Further studies would support such endeavours to ensure that the lessons withstand further evidence and are analysed properly as new case studies and theoretical developments emerge. These investigations are especially needed considering the media's 're-discovery' of and optimism for disaster diplomacy each time a disaster hits a conflict zone.

To be fair, case studies can be radically different. No guarantee exists that every future disaster-diplomacy case study must obey the same patterns as previous ones. Continuing investigations are important from research and practitioner perspectives – and for linking those perspectives for useable and used science.

As one example, disaster-diplomacy research has not yet fully engaged with the sociological literature on disaster as an agent of change, involving social and political change, as adeptly summarised by Petropoulos (2001). Many more connections have yet to be made between disaster-diplomacy research and other areas of disaster research.

An additional need is for more connections between disaster-diplomacy research and research into international affairs and international relations. Such work should connect with practitioners in those fields as well in order to understand more about what they seek and could use from the research. The analysis in this book might be too negative towards disaster diplomacy by failing to factor in these other perspectives. Alternatively, the conclusions here might be reaffirmed through wider explorations.

From a research perspective, a standard research question is 'Does a link exist between a and b?' A hypothesis can be formulated that postulates the degree or complete absence of a link and why that link does or does not exist. For disaster diplomacy, a is disaster-related activities and b is diplomacy, with the focus being a influencing b rather than a two-way relationship. Any answer to 'Does a link exist between a and b?' – which could be 'yes', 'no', 'sometimes', or most commonly 'it depends' – is important and is publishable from a research perspective.

As new case studies are analysed, the suite of possible answers may expand. Further comparison amongst different case studies and sets of cases would be feasible. A null result, stating that a case study does not display disaster diplomacy, is important too. That refutes the rising expectations in some instances that disaster diplomacy would and should be important, such as Iran–USA (section 4.2) and India–Pakistan (section 4.9). That also injects realism into cases where expectations are created that disaster diplomacy should be important, such as for vaccine diplomacy (section 4.16).

Consequently, it is important for practitioners to study disaster diplomacy and to have access to the science – which they hopefully will apply. Media and politicians can be particularly adept at jumping on a bandwagon to state that disaster-related activities should or should not lead to diplomacy, particularly after a disaster. After any major disaster in a conflict area, commentaries abound regarding the window of opportunity to create peace from the ruins or to overturn the old political order on the basis of the disaster. These changes are supposed to set the stage for reconstructing after the disaster and for reconstructing a better political system.

As the case studies show, that is not a frequent outcome in reality. Articulating that desired outcome and actively pushing for it seems frequently to close doors to more gradual change. Rather than hoping for more positive disaster-diplomacy outcomes, acting on the basis of realism might yield more long-term gains. The science needs to continually emphasise the lack of evidence supporting the hope for quick results.

This message is challenging to convey to politicians who frequently seek – and, to justify their positions, need – rapid results. Journalists, too, tend to find a better story in suggestions that disaster diplomacy will or must happen, rather than reporting pessimistic assessments. Additionally, null results are not often feasible for journalists to headline. The difficulty is demonstrated through comparing the disaster-diplomacy headline 'Peace emerges from the landslide rubble' with 'Devastating epidemic, but politics continues as usual'.

Many other tracks do not succumb to the populism to which politicians and journalists often react. The Henry Dunant Center, which in 2002 changed its name to the Center for Humanitarian Dialogue, was in Aceh in the years before the 2004 tsunami. They were building trust and encouraging dialogue between the Acehnese fighters and the Indonesian government. From a disaster-diplomacy perspective, and with the luxury of hindsight, research could indicate whether or not pursuing joint disaster-risk reduction initiatives would have been an appropriate space for the two parties to come together in order to build trust in order to address the conflict. Instead, post-tsunami reconstruction served that role.

Investigating these topics with, and explaining them for, practitioners who are aiming for change and who are willing to heed disaster-diplomacy lessons would be helpful outcomes from the science. Three main points emerge to pursue for continuing research and application.

First, why do the humanitarian imperative and disaster-risk reduction tend to be relatively low political priorities leading to little immediate success in disaster diplomacy? Second, how could the situation be proactively improved or, at minimum, how could passive regression of the situation be actively avoided? Third, under which circumstances does disaster-diplomacy work passively or could be made to work actively?

These results would be useful for practitioners involved in disaster-related activities, in diplomatic activities, and in their combination. Examples of the latter include developing high-level disaster-related agreements such as UNISDR (2005) and negotiating the delivery of disaster-risk reduction services or disaster relief. Knowing which previous case studies displayed similar characteristics to a current situation, along with the underlying reasons for people's behaviour regarding disaster diplomacy, can assist practitioners in interpreting their experiences for their own, current decision-making.

11.3 Main gaps to be overcome

In summary the main gaps to overcome regarding disaster-diplomacy research and making it useful for practice are understanding and applying that knowledge for:

- People's motivations, especially decision-makers, in disaster diplomacy case studies.
- Making disaster diplomacy work – if that is desired.
- Better linking the disaster-diplomacy work reported in this book to the vast scientific literature and practitioner work that has not yet been fully considered in the analyses.

12 The future of disaster diplomacy

The purpose of this book is to provide a scientific summary and analysis of recent disaster-diplomacy work to systematically understand and explain disaster diplomacy. That has been achieved through empirical evidence from a variety of case studies, theoretical analyses seeking an explanatory and predictive model for disaster diplomacy, and describing why such a model might not be feasible. Gaps in the current understanding of disaster diplomacy are also covered.

Disaster diplomacy tends to fail and has so many competing interests and complexities that it might not be feasible to make it work. Hope in disaster diplomacy resolving either disaster or diplomacy challenges seems to be misplaced – at least, according to the evidence of the case studies examined so far.

Despite this negative result, absence of evidence is not evidence of absence. That is, successful disaster diplomacy being absent amongst the case studies examined precludes neither (i) further case studies, either historical or in the future, showing cause for optimism in disaster diplomacy nor (ii) further depth of investigation revealing more positive outcomes or cause for more optimism based on literature or experience not fully examined here. That is, this author makes mistakes. The case cannot be closed based on one author's interpretations.

A particularly important path to pursue is to recognise that even entirely failed disaster-diplomacy attempts teach those involved about disaster diplomacy, create contacts, and generate lessons for future scenarios. As Glantz (2000) notes for the Cuba–USA case study, the high level of disaster-related scientific and technical cooperation means that when the politicians of both countries decide to strengthen ties, as happened after Fidel Castro stepped down from leading Cuba in 2008, a foundation exists for cooperation to expand.

That form of disaster diplomacy, happening at the individual level until the governmental level is ready to engage with it, could prove to be the most successful manifestation of the process. Other possible successes from individual-level disaster diplomacy are discussed by Akcinaroglu *et al.* (2008, 2011). They note that a grassroots desire for reconciliation can be a powerful driver.

Similarly, the abysmal failure of India–Pakistan disaster diplomacy following the 2001 Gujarat earthquake disaster contributed to the threat of violent conflict between the two countries later that year (Kelman, 2003). Four years later, the failure of India–Pakistan disaster diplomacy following the 2005 Kashmir earthquake was not so bitter. It was treated cautiously by the leaders, as part of the ups and downs of India–Pakistan relations (Kelman, 2006a and see also Akcinaroglu et al., 2008, 2011).

Were lessons learned from 2001 and applied, softening the 2005 situation? Greece and Turkey also learned from their disaster diplomacy roller-coaster after the 1999 earthquakes (Ker-Lindsay, 2000, 2007). That might have helped their rapprochement survive other difficult times.

This speculation requires more research to determine whether or not the negative attitude towards disaster diplomacy could be overturned, perhaps partially. Positive but more subtle outcomes might be ever-present, feeding into the slow pace of diplomacy and supporting peace-making in more understated and indirect manners than the disaster diplomacy literature to date has uncovered – or considered. The disaster-diplomacy literature might not have uncovered that because the authors have not looked for it, having become mired in the negative outlook or in short-term analyses. The possibility also exists that this discussion is false and fabricated optimism – trying to keep disaster diplomacy alive for no justifiable reason when all vital signs are long gone.

The analyses here demonstrate that disaster diplomacy is best viewed as a complex, interactive process, not as an either-or proposition where either disaster diplomacy happens or does not happen. Disaster-related activities are one factor amongst many in various forms of diplomacy, conflict reduction, and peace.

Sometimes, disaster-related factors are prominent, driving parties along or creating a space for a path towards reconciliation. Greece–Turkey and post-tsunami Aceh are examples. Sometimes, disaster-related factors are irrelevant, with a major catastrophe such as Cyclone Nargis in Burma or Hurricane Katrina in the USA failing to cause deviations in a conflict-ridden governance approach.

Is the disaster-diplomacy situation hopeless? Whether in the pursuit of disaster-related objectives or other goals, new diplomacy sometimes happens only if it is actively supported or lobbied for by parties such as political leaders, the media, popular will, or non-political heavyweights. Disaster diplomacy can be either adopted or avoided by choice.

The media are often particularly prominent in lobbying for disaster diplomacy. That can go too far, possibly to the point where expectations are raised beyond realistic outcomes, as occurred for India–Pakistan in 2001. That could also go to the point where decision-makers react adversely to suggestions on how they should act vis-à-vis disaster diplomacy, as occurred for Iran–USA after the 2003 earthquake. Rather than spurring them towards disaster-related peace, the leaders might go out of their way to avoid dealing with an enemy.

This action would show that they do not need disaster-related assistance and that they will never compromise their 'principles' by collaborating with a nemesis.

Good intentions in trying to promote disaster diplomacy can sabotage it, as could have occurred for Eritrea–Ethiopia in 2000–02. It is feasible that nothing would have brought Eritrea and Ethiopia together at that time.

Alternatively, good intentions that fail can give others ideas in trying to pursue disaster diplomacy outcomes. A good example is Greece–Turkey since 1999. Yet even where leaders choose to follow disaster diplomacy supported by the media and the grassroots, they run the risk of being rebuffed and embarrassed by the other side.

Disaster-diplomacy outcomes are never certain. Research can differentiate various disaster-diplomacy situations and influences, can explain why disaster diplomacy observations occur, and can describe when and how disaster diplomacy could do better. It cannot necessarily predict specific outcomes.

The underlying assumption is that both disaster-risk reduction and conflict reduction are the goals being sought. This assumption is a significant limitation in the analyses.

Yet this assumption is fair for many parties involved. They are interested in highlighting and trying to tackle the long-standing, fundamental, root causes of enmity and disaster vulnerability. In cases where personalities make the difference, such as Castro, the enmity might appear to be insoluble. In cases where little desire exists to avert disaster, such as not putting the needed resources into monitoring and being prepared to deal with astronomical objects that might strike the Earth, the disaster vulnerability might appear to be insoluble.

These situations never preclude trying. The efforts, spin-offs, and connections might contribute to positive disaster-diplomacy outcomes in long-term, subtle, non-linear ways – or might worsen the situation. What must be avoided is seeking quick fixes and mechanistic approaches to solving disaster, diplomacy, and disaster diplomacy challenges.

In the end, it is the choice of the parties with power regarding what they wish to see from dealing with disaster, diplomacy, and disaster diplomacy. So far, in approximately a decade of disaster diplomacy work as reported in this book, the attempts to make disaster diplomacy work have not led to direct, unambiguous success. But one decade of work emerging from principally one publication (Kelman and Koukis, 2000) does not cover everything.

However unsuccessful disaster diplomacy appears to be from the investigations presented here, options always exist to avoid sitting back passively hoping that something might somehow go right. Choices can be made to actively pursue disaster diplomacy so that it is made to work, as long as all parties involved pursue that pathway to peace.

References

Abbas, H. (2005) *Pakistan's Drift into Extremism: Allah, the Army, and America's War on Terror*, Armonk, NY: East Gate.
Abramovitz, J. (2001) *Unnatural Disasters*, Worldwatch Paper 158, Washington, DC: Worldwatch Institute.
ACNDR (2003) *Draft Civil Contingencies Bill: Comments*, Portsmouth, VA: ACNDR, (UK Advisory Committee For Natural Disaster Reduction).
AI (2005) *Amnesty International Report 2005*, London: AI (Amnesty International).
AI (2010) *Amnesty International Report 2010*, London: AI (Amnesty International).
AIDMI (2007) 'Disaster micro-insurance scheme for low-income groups', in Global Network of NGOs for Disaster Risk Reduction Building Disaster (ed.) *Building Disaster Resilient Communities: Good Practices and Lessons Learned*, Geneva: UNISDR (United Nations International Strategy for Disaster Reduction): 23–25.
Akcinaroglu, S., DiCicco, J.M. and Radziszewski, E. (2008) 'Avalanches and olive branches: natural disasters and peacemaking between interstate rivals', update of a paper prepared for the Annual Meeting of the International Studies Association, Chicago, 27 February–3 March 2007.
Akcinaroglu, S., DiCicco, J.M. and Radziszewski, E. (2011) *Avalanches and Olive Branches: A Multi-Method Analysis of Disasters and Peacemaking in Interstate Rivalries*. Political Research Quarterly, 64(2): 260–75.
Aldecoa, F. and Keating, M. (eds) (1999) *Paradiplomacy in Action: the Foreign Relations of Subnational Governments*, London: Frank Cass.
Ali, S.H. (ed.) (2007) *Peace Parks: Conservation and Conflict Resolution*, Boston: Massachusetts Institute of Technology.
Andersen, C. (1999) 'Governing aboriginal justice in Canada: constructing responsible individuals and communities through "tradition"', *Crime, Law and Social Change*, 31(4): 303–26.
Armitage, R.L. (2003) *US Policy and Iran, Testimony before the Senate Foreign Relations Committee (October 28, 2003 As Prepared)*, Washington, DC: United States Senate Foreign Relations Committee.
Aspinall, E. and Berger, M.T. (2001) 'The break-up of Indonesia? Nationalisms after decolonisation and the limits of the nation-state in post-cold war Southeast Asia', *Third World Quarterly*, 22(6): 1003–24.
Autesserre, S. (2002) 'United States "humanitarian diplomacy" in South Sudan', *Journal of Humanitarian Assistance*, March 2002: online. Available: www.jha.ac/articles/a085.htm (accessed 4 May 2009).

Balamir, M. (2005) 'Ways of understanding urban earthquake risks', *Book of Abstracts* from 'Rethinking Inequalities' 7th Conference of European Sociological Association, Institute of Sociology Nicolaus Copernicus University of Torun, Poland: 132.

Baldacchino, G. (2004) 'Autonomous but not sovereign? A review of island sub-nationalism', *Canadian Review of Studies in Nationalism*, 31(1–2): 77–91.

Baldacchino, G. and Milne, D. (eds) (2000) *Lessons from the Political Economy of Small Islands: The Resourcefulness of Jurisdiction*, New York City: St. Martin's Press.

Baldacchino, G. and Milne, D. (eds) (2009) *The Case for Non-Sovereignty: Lessons from Sub-National Island Jurisdictions*, London: Taylor & Francis.

Barrett, S. (2003) 'Diplomacy vs disease: the history of smallpox eradication shows how diplomatic strategies could help tackle worldwide infectious disease', *SAISPHERE*, 2003: 25–27.

Bartmann, B. (2006) 'In or out: sub-national island jurisdictions and the antechamber of para-diplomacy', *The Round Table*, 95(386): 541–59.

Baskin, C. (2002) 'Holistic healing and accountability: indigenous restorative justice', *Child Care in Practice*, 8(2): 133–36.

Bayne, N. and Woodcock, S. (eds) (2007) *The New Economic Diplomacy: Decision-making and Negotiation in International Economic Relations*, Aldershot: Ashgate Publishing.

Beeson, M. and Higgott, R. (2005) 'Institutionalism and US foreign policy: theory and practice in comparative historical perspective', *Third World Quarterly*, 26(7): 1173–88.

Bender, G.J., Coleman, J.S. and Sklar, R.L. (1985), *African Crisis Areas and U.S. Foreign Policy*, Berkeley and Los Angeles: University of California Press.

Benedick, R.E. (1998) *Ozone Diplomacy: New Directions in Safeguarding the Planet*, Cambridge, MA: Harvard University Press.

Bennett, J. (1999) *North Korea: The Politics of Food Aid*. RRN (Relief and Rehabilitation Network) Network Paper 28, London: Overseas Development Institute.

Bereciartu, G.J. (translated by W.A. Douglass) (1994) *Decline of the Nation-State*, Reno, NV: University of Nevada Press.

Biddick, T.V. (1989) 'Diplomatic rivalry in the South Pacific: the PRC and Taiwan', *Asian Survey*, 29(8): 800–815.

Bird, D.K. (2009) 'The use of questionnaires for acquiring information on public perception of natural hazards and risk mitigation – a review of current knowledge and practice', *Natural Hazards and Earth Systems Sciences*, 9: 1307–25.

Blum, S. (2003) 'Chinese views of US hegemony', *Journal of Contemporary China*, 12(35): 239–64.

Bolt, B.A. (1993) *Earthquakes (Newly Revised and Expanded)*, New York: W.H. Freeman and Company.

Bose, S. (2003) *Kashmir: Roots of Conflict, Paths to Peace*, Cambridge, MA: Harvard University Press.

Bosher, L.S. (ed.) (2008) *Hazards and the Built Environment: Attaining Built-in Resilience*, London: Taylor & Francis.

Box, G.E.P. and Draper, N.R. (1987) *Empirical Model-Building and Response Surfaces*, New York: John Wiley and Sons.

Brancati, D. (2007) 'Political aftershocks: the impact of earthquakes on intrastate conflict', *Journal of Conflict Resolution*, 51(5): 715–43.

Broder, S., Hoffman, S.L. and Hotez, P.J. (2002) 'Cures for the Third World's problems', *EMBO Reports*, 3(9): 806–12.

Bryant, E. (2008) *Tsunami: The Underrated Hazard*, Berlin: Springer.

Buhaug, H. (2010a) 'Climate not to blame for African civil wars', *Proceedings of the National Academy of Sciences*, 107(38): 16477–82.

Buhaug, H. (2010b) 'Reply to Burke *et al.*: bias and climate war research', *Proceedings of the National Academy of Sciences*, 107(51): E186–E187.

Buhaug, H., Gleditsch, N.P. and Theisen, O.M. (2008) 'Implications of climate change for armed conflict', paper presented at Social Dimensions of Climate Change, Washington, DC, The World Bank, 5–6 March.

Buhaug, H., Gleditsch, N.P. and Theisen, O.M. (2010) 'Implications of climate change for armed conflict', in R. Mearns and A. Norton (eds), *Social Dimensions of Climate Change: Equity and Vulnerability in a Warming World*, Washington, DC: The World Bank: 75–101.

Bull, H. (1977) *The Anarchical Society: A Study of Order in World Politics*, New York: Columbia University Press.

Bullion, A.J. (1995) *India, Sri Lanka and the Tamil Crisis, 1976–1994: an international perspective*, London: Pinter.

Burke, M.B., Miguel, E., Satyanath, S., Dykemae, J.A. and Lobell, D.B. (2009) 'Warming increases the risk of civil war in Africa', *Proceedings of the National Academy of Sciences*, 106(49): 20670–74.

Burke, M.B., Miguel, E., Satyanath, S., Dykemae, J.A. and Lobell, D.B. (2010a) 'Climate robustly linked to African civil war', *Proceedings of the National Academy of Sciences*, 107(51): E185.

Burke, M.B., Miguel, E., Satyanath, S., Dykemae, J.A. and Lobell, D.B. (2010b) 'Reply to Sutton *et al.*: Relationship between temperature and conflict is robust', *Proceedings of the National Academy of Sciences*, 107(25): E103.

Bush, G.W. 2002 (29 January) *The President's State of the Union Address*, Washington, DC: Office of the Press Secretary, U.S. Government.

Butler, R.A. and Laurance, W.F. (2008) 'New strategies for conserving tropical forests', *Trends in Ecology & Evolution*, 23(9): 469–72.

Camroux, D. and Okfen, N. (2004) 'Introduction: 9/11 and US–Asian relations: towards a new "New World Order"?', *The Pacific Review*, 17(2): 163–77.

Caplan, A. (2009) 'Is disease eradication ethical?', *The Lancet*, 373(9682): 2192–93.

Carroll, M.S., Higgins, L.L., Cohn, P.J. and Burchfield, J. (2006) 'Community Wildfire Events as a Source of Social Conflict', *Rural Sociology*, 71(2): 261–80.

Carusi, A., Perozzib, E. and Scholl, H. (2005) 'Mitigation strategy', *Comptes Rendus Physique*, 6(3): 367–74.

Cater, N. (2003) 'Viewpoint: When disaster opens the door to dialogue', *AlertNet*, 30 December 2003: online (no longer available via AlertNet). Available: www.disasterdiplomacy.org/iranusa.html (accessed 10 July 2010).

Cellino, A., Somma, R., Tommasi, L., Paolinetti, R., Muinonen, K., Virtanen, J., Tedesco, E.F. and Delbò, M. (2006) 'NERO: general concept of a near-earth object radiometric observatory', *Advances in Space Research*, 37: 153–60.

Chasek, P. (2001) *Earth Negotiations: Analyzing Thirty Years of Environmental Diplomacy*, Tokyo: UNU Press.

Chattopadhyaya, H.P. (1994) *Ethnic Unrest in Modern Sri Lanka: an Account of Tamil–Sinhalese Race Relations*, New Delhi, MD Publications.

Chen, R. (2003) 'China Perceives America: perspectives of international relations experts', *Journal of Contemporary China*, 12(35): 285–97.

CIA. (2005) 'Indonesia', *CIA World Fact Book*, Washington, DC: CIA (Central Intelligence Agency).

Clay, E. (1999) *An Evaluation of HMG's Response to the Montserrat Volcanic Emergency*, DFID (Department for International Development) Evaluation Report EV635, London: DFID.
Clifford, R.A. (1956) *The Rio Grande Flood; a Comparative Study of Border Communities in Disaster*, Publication No. 458, Washington, DC: National Academy of Sciences – National Research Council.
Coburn, A., Hughes, R., Spence, R. and Pomonis, A. (1995) *Technical Principles of Building for Safety*, London: Intermediate Technology Limited.
Collier, P. (2000) 'Rebellion as a quasi-criminal activity', *Journal of Conflict Resolution*, 44(6): 839–53.
Collier, P. and Sambanis, N. (eds) (2005) *Understanding civil war: evidence and analysis. Volume I: Africa*, Washington, DC: The World Bank.
Comfort, L. (2000) 'Disaster: agent of diplomacy or change in international affairs?', *Cambridge Review of International Affairs*, XIV(1): 277–94.
Connell, J. (1997) *Papua New Guinea: The Struggle for Development*, London: Routledge.
Cotton, W.R. and Pielke, Sr., R.A. (1995) *Human Impacts on Weather and Climate*. Cambridge: Cambridge University Press.
Croall, J. (1997) *Lets Act Locally: The Growth of Local Exchange Trading Systems*, London: Calouste Gulbenkian.
Crowther D., Greene, A-M. and Hosking, D.M. (2002) 'Local economic trading schemes and their implications for marketing assumptions, concepts and practices', *Management Decision*, 40(4): 354–62.
Cuny, F. (1983) *Disasters and Development*, Oxford: Oxford University Press.
Cyranoski, D. (2005) 'Get off the beach – now!' *Nature*, 433: 354.
Dallaire, R. (2003) *Shake Hands with the Devil: The Failure of Humanity in Rwanda*, New York: Carroll & Graf.
Davidson, W.D. and Montville, J.V. (1981) 'Foreign policy according to Freud', *Foreign Affairs*, Winter 1981–82: 145–57.
De Boer, J.Z. and Sanders, D.T. (2004) *Volcanoes in Human History: The Far-Reaching Effects of Major Eruptions*, Princeton, NJ: Princeton University Press.
De Boer, J.Z. and Sanders, D.T. (2005) *Earthquakes in Human History: The Far-Reaching Effects of Seismic Disruptions*, Princeton, NJ: Princeton University Press, Princeton.
Der Derian, J. (1987) *On Diplomacy: a Genealogy of Western Estrangement*, Oxford: Blackwell.
DHS (2006) *A Performance Review of FEMA's Disaster Management Activities in Response to Hurricane Katrina*, Washington, DC: DHS (Department of Homeland Security), Office of Inspector General, Office of Inspections and Special Reviews.
Diamond, L. and McDonald, J. (1993) *Multi-Track Diplomacy: A Systems Approach to Peace*, Washington, DC: Institute for Multi-Track Diplomacy.
Dillon, M. (1997) 'Ethnic, religious and political conflict on China's northwestern borders: the background to the violence in Xinjiang', *IBRU Boundary and Security Bulletin*, Spring: 80–86.
Dixit, A. (2003) 'The community based program of NSET for earthquake disaster management', paper presented at the International Conference on Total Disaster Risk Management, Kobe, 2–4 December.
Dixit, J.N. (2002) *India-Pakistan in War and Peace*, London: Routledge.
Dobriansky, P.J. (2005) 'Responding to the global threat of Avian and pandemic influenza', *Remarks to the Senate Committee on Foreign Relations*, 9 November, Washington, DC: Senate Committee on Foreign Relations.

Dove, M.R. (1998) 'Local dimensions of 'global' environmental debates'., in A. Kalland and G. Persoon (eds), *Environmental Movements in Asia*, Richmond, Surrey: Curzon Press: 44–64.

Dove, M.R. and Khan, M.H. (1995) 'Competing constructions of calamity: the April 1991 Bangladesh cyclone', *Population and Environment*, 16(5): 445–71.

Doyle, M.W. (1983a) 'Kant, liberal legacies, and foreign affairs', *Philosophy and Public Affairs*, 12(3): 205–35.

Doyle, M.W. (1983b) 'Kant, liberal legacies, and foreign affairs, part 2', *Philosophy and Public Affairs*, 12(4): 323–53.

Drury, A.C. and Olson, R.S. (1998) 'Disasters and political unrest: an empirical investigation', *Journal of Contingencies and Crisis Management*, 6(3): 153–61.

Duchacek, I.D., Latouch, D. and Stevenson, G. (eds) (1988) *Perforated Sovereignties and International Relations: Trans-Sovereign Contacts of sub-national Governments*, Westport, CT: Greenwood Press.

Duffy, R. (2001) 'Peace parks: the paradox of globalisation', *Geopolitics*, 6(2): 1–26.

Dunn, D.H. (2003) 'Myths, motivations and "misunderestimations": the Bush administration and Iraq', *International Affairs*, 79(2): 279–97.

Dunn, D.H. (2006) '"Quacking like a duck"? Bush II and presidential power in the second term', *International Affairs*, 82(1): 95–120.

Dunne, M. (2003) 'The United States, the United Nations and Iraq: "multilateralism of a kind"', *International Affairs*, 79(2): 257–77.

Economides, S. (2005) 'The Europeanisation of Greek foreign policy', *West European Politics*, 28(2): 471–91.

'Elián González by and through Lazaro González v. Janet Reno et al.' (2000) United States Court of Appeals for the Eleventh Circuit, no. 00–11424, DC Docket No. 00–206-CV-KMM (23 June).

Enia, J. (2008) 'Peace in its Wake? The 2004 Tsunami and internal conflict in Indonesia and Sri Lanka', *Journal of Public and International Affairs*, 19: 7–27.

Enia, J. (2009) 'Shaking the foundations of violent civil conflict: institutions, disasters, and the political economies of state-rebel interaction', unpublished PhD dissertation, University of Southern California.

'Exchange' (2003) 'Thomas Homer-Dixon, Nancy Peluso, and Michael Watts on violent environments', *ECSP Report*, 9: 89–96.

FAO (2000) A human catastrophe looms in the Horn of Africa. *FAO Global Information and Early Warning System on Food and Agriculture*, Special Alert no. 306, Rome: FAO (Food and Agriculture Organization of the United Nations).

Farer, T. (1971) 'Humanitarian law and armed conflicts: toward the definition of "international armed conflict"', *Columbia Law Review*, 71(1): 37–72.

Fidler, D.P. (2001) 'The globalization of public health: the first 100 years of international health diplomacy', *Bulletin of the World Health Organization.*, 79(9): 842–49.

Fischetti, M. (2001) 'Drowning New Orleans', *Scientific American*, October: 76–85.

Fortes, M. (2002) 'The Spratlys as a zone of peace: the transboundary biosphere reserve concept at work', *Wise Coastal Practices for Sustainable Human Development Forum*: online. Available: www.csiwisepractices.org/?read=411 (accessed 1 February 2004).

Furlong, G.T. (2005) 'The conflict resolution toolbox: models & maps for analyzing, diagnosing and resolving conflict', Mississauga, ON: John Wiley and Sons Canada.

Gaillard, J.-C., Clavé, E., and Kelman, I. (2008) 'Wave of peace? Tsunami disaster diplomacy in Aceh, Indonesia', *Geoforum*, 39(1): 511–26.

Gaillard, J.-C., Kelman, I. and Orillos, M.F. (2009) 'US–Philippines military relations after the Mt Pinatubo eruption in 1991: a disaster diplomacy perspective', *European Journal of East Asian Studies*, 8(2): 301–30.

Gaillard, J.-C., Liamzon, C.C. and Villanueva, J.D. (2007) '"Natural" disaster? A retrospect into the causes of the late-2004 typhoon disaster in Eastern Luzon, Philippines', *Environmental hazards*, 7(4): 257–70.

Galtung, J. (1997) *Health as a Bridge for Peace in the Context of Humanitarian Action in Complex Emergency Situations*, Geneva: Division of Emergency and Humanitarian Action, World Health Organization.

Ganapti, E., Kelman, I. and Koukis, T. (2010) 'Analyzing Greek-Turkish Disaster-Related Cooperation: A Disaster Diplomacy Perspective', *Cooperation and Conflict*, 45(2): 162–85.

GAO (2003) *Alaska Native Villages: Most are Affected by Flooding and Erosion but Few Qualify for Federal Assistance*, Washington, DC: GAO (United States General Accounting Office).

GAO (2006) *Hurricane Katrina: Comprehensive Policies and Procedures Are Needed to Ensure Appropriate Use of and Accountability for International Assistance*, Washington, DC: United States GAO (Government Accountability Office).

George, J. (2005) 'Leo Strauss, neoconservatism and us foreign policy: esoteric nihilism and the bush doctrine', *International Politics*, 42(2): 174–202.

GHA. (2009) *GHA Report 2009*, Wells: GHA (Global Humanitarian Assistance) and Development Initiative.

GHI. (2006) *Seismic Safety for Adobe Homes: What Everyone Should Know*, Palo Alto: GHI (GeoHazards International).

Gibbs, A. and King, D. (2002) 'Alternatives to custody in the new zealand criminal justice system: current features and future prospects', *Social Policy and Administration*, 36(4): 392–407.

Glantz, M.H. (1976) *The Politics of Natural Disaster: The Case of the Sahel Drought*, New York: Praeger.

Glantz, M.H. (1994a) 'Creeping environmental problems', *The World & I*, June: 218–25.

Glantz, M.H. (1994b) 'Creeping environmental phenomena: Are societies equipped to deal with them?', in M.H. Glantz (ed.) *Creeping Environmental Phenomena and Societal Responses to Them, Proceedings of Workshop held 7–10 February 1994 in Boulder, Colorado*, Boulder, CO: NCAR/ESIG: 1–10.

Glantz, M.H. (ed) (1997) 'Using science against famine: food security, famine early warning, and El Niño' *Internet Journal of African Studies*, 1(2): online. Available: http://ccb.colorado.edu/ijas/ijasno2/ijasno2.html (accessed 23 November 2010).

Glantz, M.H. (ed) (1999) *Creeping Environmental Problems and Sustainable Development in the Aral Sea Basin*, Cambridge: Cambridge University Press.

Glantz, M.H. (2000) 'Climate-related disaster diplomacy: a US–Cuban case study', *Cambridge Review of International Affairs*, XIV(1): 233–53.

Glantz, M.H. (2003) *Climate Affairs: A Primer*, Washington, DC: Island Press.

Glantz, M.H. (2008) 'Hurricane Katrina as a "teachable moment"', *Advances in Geosciences*, 14: 287–94.

Glantz, M.H. and Katz, R.W. (1977) 'When is a drought a drought?', *Nature* 267: 192–93.

Gleditsch, N.P., Wallensteen, P., Eriksson, M., Sollenberg, M. and Strand, H. (2002) 'Armed conflict 1946–2001: a new dataset', *Journal of Peace Research*, 39(5): 615–37.

Goldblat, J. (1975) 'The prohibition of environmental warfare', *Ambio*, 4(5/6): 186–90.
Gonzalez, E. and Ronfeldt, D. (1986) *Castro, Cuba, and the World*, Santa Monica, CA: Rand.
Goren, S. (2001) 'Healing the victim, the young offender, and the community via restorative justice: an international perspective', *Issues in Mental Health Nursing*, 22(2): 137–49.
Harbom, L. and Wallensteen, P. (2009) 'Armed conflict, 1946–2008', *Journal of Peace Research*, 46(4): 477–87.
Hartmann, B. (2010) 'Rethinking climate refugees and climate conflict: rhetoric, reality, and the politics of policy discourse', *Journal of International Development*, 22(1): 233–46.
He, B. and Reid, A. (2004) 'Special issue editors' introduction: four approaches to the Aceh question Source', *Asian Ethnicity*, 5(3): 293–300.
Heath, O. (1999) 'Anatomy of bjp's rise to power: social, regional and political expansion in 1990s', *Economic and Political Weekly*, 34(34/35): 2511–17.
Henderson, D.A. (1999) 'Eradication: lessons from the past', *MMWR (Morbidity and Mortality Weekly Report)*, 48(1): 16–22.
Hewitt, K. (ed.) (1983) *Interpretations of Calamity from the Viewpoint of Human Ecology*, London: Allen & Unwin.
Hewitt, K. (1997) *Regions of Risk: A Geographical Introduction to Disasters*, Harlow, Essex: Addison Wesley Longman.
Hoffman, S. (2003) 'The hidden victims of disaster', *Environmental Hazards*, 5(3–4): 67–70.
Holland, J.H. (1995) *Hidden Order: How Adaptation Builds Complexity*, Reading, MA: Addison Wesley Publishing Company.
Holloway, A. (2000) 'Drought emergency, yes … drought disaster, no: Southern Africa 1991–93', *Cambridge Review of International Affairs*, XIV(1): 254–76.
Holloway, A. (2009) 'Crafting disaster risk science: environmental and geographical science sans frontières', *Gateways: International Journal of Community Research and Engagement*. 2: 98–118.
Homer-Dixon, T. (1999) *Environment, Scarcity and Violence*, Princeton, NJ: Princeton University Press.
Hopkins, D.R. and Ruiz-Tiben, E. (1991) 'Strategies for dracunculiasis eradication', *Bulletin of the World Health Organization*, 69(5): 533–40.
Hotez, P.J. (2001a) 'Vaccine diplomacy', *Foreign Policy*, May/June: 68–69.
Hotez, P.J. (2001b) 'Vaccines as instruments of foreign policy', *EMBO Reports*, 2(10): 862–68.
Hotez, P.J. (2004) 'The promise of medical science and biotechnology for North Korea and the relevance of U.S. "vaccine diplomacy"', *Korea Society Quarterly*, 13(4): 15–18.
Hotez, P.J. (2010) 'Peace through vaccine diplomacy', *Science*, 327(5971): 1301.
Hotez, P.J. and Thompson, T.G. (2009) 'Waging peace through neglected tropical disease control: a US foreign policy for the bottom billion', *PLoS Neglected Tropical Diseases*, 3(1): e346.
House of Commons (2004) *Civil Contingencies Bill*, London: House of Commons.
HRW (2005) *World Report 2005: Events of 2004*, New York: HRW (Human Rights Watch).
HRW (2009) *New Castro, Same Cuba: Political Prisoners in the Post-Fidel Era*, New York: HRW (Human Rights Watch).

Hurst, S. (2005) 'Myths of neoconservatism: George W. Bush's "neo-conservative" foreign policy revisited', *International Politics*, 42: 75–96.

IEA (2009) *CO_2 Emissions from Fuel Combustion*, Paris: IEA (International Energy Agency).

IFRC (2000) *World Disasters Report 2000: Focus on Public Health*, Geneva: IFRC (International Federation of Red Cross and Red Crescent Societies).

Ihonvbere, J.O. (1996) 'Where Is the Third Wave? A critical evaluation of africa's non-transition to democracy', *Africa Today*, 43(4): 343–67.

IPCC (2007) *IPCC Fourth Assessment Report*, Geneva: IPCC (Intergovernmental Panel on Climate Change).

Jayaraman, K.S. (2005) 'India makes waves over tsunami warning system', *Nature*, 438: 1060–61.

Jonkman, S.N. (2005) 'Global perspectives of loss of human life caused by floods', *Natural Hazards*, 34(2): 151–75.

Jonkman, S.N. and Kelman, I. (2005) 'An analysis of causes and circumstances of flood disaster deaths', *Disasters*, 29(1): 75–97.

Kahan, A.M., Rottner, D., Sena, R. and Keyes Jr., C.G. (eds) (1995) *Guidelines for Cloud Seeding to Augment Precipitation*, Reston, VA: American Society of Civil Engineers.

Kahl, C. (2002) 'Violent environments, Nancy Lee Peluso and Michael Watts (Eds.) (Review)', *ECSP Report*, 8: 135–43.

Kapur, D. (2005) 'Remittances: the new development mantra?', in S.M. Maimbo and D. Ratha (eds), *Remittances: Development Impact and Future Prospects*, Washington, DC: World Bank: 331–60.

Kardol, R. (1999) 'Proposed inhabited artificial islands in international waters: international law analysis in regards to resource use, law of the sea and norms of self-determination and state recognition', unpublished master's thesis, Universiteit van Amsterdam.

Karp, E., Sebbag, G., Peiser, J., Dukhno, O., Ovnat, A., Levy, I., Hyam, E., Blumenfeld, A., Kluger, Y., Simon, D. and Shaked, G. (2007) 'Mass casualty incident after the Taba terrorist attack: an organisational and medical challenge', *Disasters*, 31(1): 104–12.

Kelman, I. (2003) 'Beyond disaster, beyond diplomacy', in M. Pelling (ed.) *Natural Disasters and Development in a Globalizing World*, London: Routledge: 110–23.

Kelman, I. (2005) 'Tsunami diplomacy: will the 26 December, 2004 tsunami bring peace to the affected countries?', *Sociological Research Online*, 10(1): online. Available: www.socresonline.org.uk/10/1/kelman.html (accessed 12 June 2007).

Kelman, I. (2006a) 'Acting on disaster diplomacy', *Journal of International Affairs*, 59(2): 215–40.

Kelman, I. (2006b) 'Island security and disaster diplomacy in the context of climate Change', *Les Cahiers de la Sécurité*, 63: 61–94.

Kelman, I. (2007) 'Hurricane Katrina disaster diplomacy', *Disasters*, 31(3): 288–309.

Kelman, I. (2010) 'Tying disaster diplomacy in knots', in G.T. Overton (ed.) *Foreign Policy in an Interconnected World*, Hauppauge, NY: Nova Publishers: 59–73.

Kelman, I. and Conrich, B. (2011) 'A framework for island disaster para-diplomacy', in S.R. Sensarma and A. Sarkar (eds) *Disaster Risk Management: Local, Regional and International Level Cooperation*, New Delhi: Concept Publishing, New Delhi: forthcoming.

Kelman, I., Davies, M., Mitchell, T., Orr, I. and Conrich, B. (2006) 'Island Disaster Para-Diplomacy in the Commonwealth', *The Round Table: The Commonwealth Journal of International Affairs*, 95(386): 561–74.
Kelman, I. and Koukis, T. (eds) (2000) 'Disaster diplomacy', special section in *Cambridge Review of International Affairs*, XIV(1): 214–94.
Kent, G. (2005) *Freedom from Want: The Human Right to Adequate Food*, Washington, DC: Georgetown University Press.
Keridis, D. (2006) 'Earthquakes, diplomacy, and new thinking in foreign policy', *Fletcher Forum Of World Affairs*, 30(1): 207–14.
Ker-Lindsay, J. (2000) 'Greek–Turkish rapprochement: the impact of "disaster diplomacy"?', *Cambridge Review of International Affairs*, XIV(1): 215–32.
Ker-Lindsay, J. (2007) *Crisis and Conciliation: A Year of Rapprochement between Greece and Turkey*, London: IB Tauris.
Kim, S.C., Chen, J., Park, K. and Choi, J.K. (1998) 'Coastal surges from extratropical storms on the west coast of the Korean Peninsula', *Journal of Coastal Research*, 14(2): 660–66.
Kouchner, B. (2008) 'Burma', *Le Monde*, 20 May.
Kronstadt, K.A. (2004) *CRS Issue Brief for Congress: India–U.S. Relations*, Congressional Research Service, Washington, DC: The Library of Congress.
Kwa, C. (2001) 'The rise and fall of weather modification: changes in American attitudes toward technology, nature, and society', in C.A. Miller and P.N. Edwards (eds) *Changing the Atmosphere: Expert Knowledge and Environmental Governance*, Boston: Massachusetts Institute of Technology: 135–65.
Lacina, B.A. and Gleditsch, N.P. (2005) 'Monitoring trends in global combat: a new dataset of battle deaths', *European Journal of Population*, 21(2–3): 145–66.
Lake, D.A. (2006) 'American hegemony and the future of East–West relations', *International Studies Perspectives*, 7: 23–30.
Landau, D. and Landau, S. (1997) 'Confidence-building measures in mediation', *Mediation Quarterly*, 15(2): 97–103.
La Red, OSSO, and ISDR (2002) *Comparative Analysis of Disaster Databases: Final Report*, Panama City and Geneva: La Red and OSSO for UNDP and ISDR.
Laws of Tonga (1988) *Criminal Offences*, Nuku'Alofa: Government of Tonga.
'Lazaro Gonzalez v. Janet Reno et al.' (2000) 'Certiorari: petition concerning Elian Gonzalez denied (99–2079)', *Journal of the Supreme Court of the United States*, October Term 1999 (28 June): 1020.
Le Billon, P. and Waizenegger, A. (2007) 'Peace in the wake of disaster? secessionist conflicts and the 2004 Indian Ocean tsunami', *Transactions of the Institute of British Geographers*, 32(3): 411–27.
Lecours, A. (2002) 'Paradiplomacy: reflections on the foreign policy and international relations of regions', *International Negotiation*, 7: 91–114.
Leffler, M.P. (2005) '9/11 and American foreign policy', *Diplomatic History*, 29(3): 395–413.
Lewis, J. (1984) 'Environmental interpretations of natural disaster mitigation: the crucial need', *The Environmentalist*, 4: 177–80.
Lewis, J. (1988) 'On the line: an open letter in response to "confronting natural disasters, an international decade for natural hazard reduction"', *Natural Hazards Observer*, XII(4): 4.
Lewis, J. (1999) *Development in Disaster-prone Places: Studies of Vulnerability*, London: Intermediate Technology Publications.

Lewis, J. (2003) 'Housing construction in earthquake-prone places: perspectives, priorities and projections for development', *Australian Journal of Emergency Management*, 18(2): 35–44.

Lewis, J. (2007) 'Inverse disaster diplomacy', Natural-Hazards-Disasters Jiscmail listserv. Online posting. Available www.jiscmail.ac.uk/cgi-bin/webadmin?A2=ind0705&L= NATURAL-HAZARDS-DISASTERS&F=&S=&P=5346 (accessed 9 May 2007).

Lortan, F. (2000) 'The Ethiopia–Eritrea conflict: a fragile peace', *African Security Review*, 9(4): online. Available: www.iss.co.za/pubs/ASR/9No4/Lortan.html (accessed 4 December 2010).

Malfatto, P.-L. and Vallet, E. (2004) 'Water geopolitics in North America', paper presented at the Annual Meeting of the International Studies Association, Montreal, 17 March.

Mandel, R. (2002) 'Security and natural disasters', *Journal of Conflict Studies*, XXXII (Fall): 118–43.

Mardhatillah, F. (2004) 'Ethno-methodological reflections on the case of Acehnese conflict', in L. Trijono (ed.) *The Making of Ethnic and Religious Conflicts in Southeast Asia: Cases and Resolutions*, Yogyakarta: CSPS Books: 205–29.

Markovits, C., Pouchepadass, J. and Subrahmanyam, S. (eds) (2003) *Society and Circulation: Mobile People and Itinerant Cultures in South Asia, 1750–1950*, New Delhi, Permanent Black.

Martin, B. (1979) *The Bias of Science*, Canberra: Society for Social Responsibility in Science.

Mavrogenis, S. (2009) *Disaster Diplomacy and Politics. The Case Study of the 1999 Greece–Turkey earthquakes*, unpublished master's dissertation, Kings College London.

Maxwell, D. (2002) 'Why do famines persist? A brief review of Ethiopia 1999–2000', *IDS Bulletin*, 33(4): 48–54.

McCorkindale, L. (1994) 'Food aid: human right or weapon of war?', *British Food Journal*, 96(3): 5–11.

McNamara, K.E. and Gibson, C. (2009) '"We do not want to leave our land": Pacific ambassadors at the United Nations resist the category of "climate refugees"', *Geoforum*, 40: 475–83.

Menzel, D.C. (2006) 'The Katrina aftermath: a failure of federalism or leadership?', *Public Administration Review*, 66(6): 808–12.

Mileti, D. and 136 contributing authors (1999) *Disasters by Design: A Reassessment of Natural Hazards in the United States*, Washington, DC: Joseph Henry Press.

Miller, J. and Kenedi, A. (eds) (2003) *Inside Cuba: the History, Culture, and Politics of an Outlaw Nation*, New York: Marlowe and Company.

MMWR (2006) 'Resurgence of wild poliovirus type 1 transmission and consequences of importation – 21 countries, 2002–5', *MMWR (Morbidity and Mortality Weekly Report)*, 55(6): 145–50.

Mohan, A. (1992) 'Historical roots of the Kashmir conflict', *Studies in Conflict and Terrorism*, 15(4): 283–308.

Montevideo Convention (1933) *Montevideo Convention on the Rights and Duties of States*, published 26 December, Montevideo.

Morfit, M. (2006) 'Staying on the road to Helsinki: why the Aceh agreement was possible in August 2005', paper presented at Building Permanent Peace in Aceh: One Year after the Helsinki Accord, Jakarta, 14 August.

Munich Re (2006) *Topics Geo: Significant Natural Catastrophes in 2005*, Munich: Munich Re Group.

Naranjo Diaz, L. (2003) 'Hurricane early warning in Cuba: an uncommon experience', in M.H. Glantz (ed.) *Usable Science 8: Early Warning Systems: Do's and Don'ts*, Report from the workshop held 20–23 October in Shanghai, China, Boulder, CO: National Center for Atmospheric Research: 59–63.

National Civil Defence Emergency Management Plan Order (2005) Wellington: Government of New Zealand.

NDMC (2006) *What is Drought?* Lincoln, NE: NDMC (National Drought Mitigation Center).

Nel, P. and Righarts, M. (2008) 'Natural disasters and the risk of violent civil conflict', *International Studies Quarterly*, 52: 159–85.

Nelson, T. (2010a) 'Rejecting the gift horse: international politics of disaster aid refusal', *Conflict, Security & Development*, 10(3): 379–402.

Nelson, T. (2010b) 'When disaster strikes: on the relationship between natural disaster and interstate conflict', *Global Change, Peace & Security*, 22(2): 155–74.

Noah, D.L., Huebner, K.D., Darling, R.G. and Waeckerle, J.F. (2002) 'The history and threat of biological warfare and terrorism', *Emergency Medicine Clinics of North America*, 20(2): 255–71.

Nordås, R. and Gleditsch, N.P. (2009) 'IPCC and The climate–conflict nexus', paper presented at the 50th Convention of the International Studies Association, New York, 15–18 February.

Norris, J. (2007) *The Disaster Gypsies: Humanitarian Workers in the World's Deadliest Conflicts*, Westport, CT: Praeger Security International.

Obama, B.H. (2009) *Nobel Lecture by Barack H. Obama*, Oslo, 10 December.

OECD (2008) *Aid Targets Slipping Out of Reach?* Paris: OECD (Organisation for Economic Co-operation and Development).

OEP (2000) *State of Louisiana Emergency Operations Plan. Supplement 1A: Southeast Louisiana Hurricane Evacuation And Sheltering Plan (Revised January 2000)*, Baton Rouge, LO: Louisiana Office of Homeland Security and Emergency Preparedness.

Ogawa, Y., Fernandez, A.L. and Yoshimura, T. (2005) 'Town watching as a tool for citizen participation in developing countries: applications in disaster training', *International Journal of Mass Emergencies and Disasters*, 23(2): 5–36.

O'Keefe, P., Westgate, K. and Wisner, B. (1976) 'Taking the naturalness out of natural disasters', *Nature*, 260: 566–67.

Oliver-Smith, T. (1986) *The Martyred City: Death and Rebirth in the Andes*, Albuquerque, NM: University of New Mexico Press.

Olson, R.S. and Drury, A.C. (1997) 'Un-therapeutic communities: a cross-national analysis of post-disaster political unrest', *International Journal of Mass Emergencies and Disasters*, 15(2): 221–38.

Olson, R.S. and Gawronski, V.T. (2010) 'From disaster event to political crisis: a "5C+A" framework for analysis', *International Studies Perspectives*, 11: 205–21.

OMB (2005) *Budget of the United States Government Fiscal Year 2006*, OMB (Office of Management and Budget), Washington, DC: US Government.

Pacione M. (1997) 'Local exchange trading systems – a rural response to the globalization of capitalism?', *Journal of Rural Studies*, 13(4): 415–27.

Parmer, I. (2005) 'Catalysing events, think tanks and american foreign policy shifts: a comparative analysis of the impacts of Pearl Harbor 1941 and 11 September 2001', *Government and Opposition*, 40(1): 1–25.

Pattullo, P. (2000) *Fire from the Mountain: The Tragedy of Montserrat and the Betrayal of its People*, London: Constable & Robinson.

Peduzzi, P., Dao, H. and Herold, C. (2005) 'Mapping disastrous natural hazards using global datasets', *Natural Hazards*, 35: 265–89.

Peluso, N. and Watts, M. (eds) (2001) *Violent Environments*, Ithaca, NY: Cornell University Press.

Perrow, C. (2005) 'Using Organizations: The Case of FEMA', *Homeland Security Affairs*, 1(2): article 4, 1–8.

Peter, N., Bartona, A., Robinson, D. and Salotti, J.-M. (2004) 'Charting response options for threatening near-earth objects', *Acta Astronautica*, 55(3–9): 325–34.

Petropoulos, N.P. (2001) 'The impact of the August–September 1999 earthquakes on Greco-Turkish relations: an exploratory study', paper presented at the 5th Conference of the European Sociological Association, Helsinki, 28 August–1 September.

Platt, R.H. (1999) *Disasters and Democracy: The Politics of Extreme Natural Events*, Washington, DC: Island Press.

Posner, R.A. (2001) *Breaking the Deadlock: the 2000 Election, the Constitution, and the Courts*, Princeton, NJ: Princeton University Press.

Price, S.D. and Egan, M.P. (2001) 'Space based infrared detection and characterization of near earth objects', *Advanced Space Research*, 28(8): 1117–27.

Quarantelli, E.L. (2001) 'Statistical and conceptual problems in the study of disasters', *Disaster Prevention and Management*, 10(5): 325–38.

Quarantelli, E.L. and Dynes, R.R. (1976) 'Community conflict: its absence and presence in natural disasters', *Mass Emergencies*, 1: 139–52.

Rajagopalan, S. (2006) 'Silver linings: natural disasters, international relations and political change in South Asia, 2004–5', *Defense & Security Analysis*, 22(4): 451–68.

Rajasingham-Senanayake, D. (2005) 'Sri Lanka and the violence of reconstruction', *Development*, 48, 111–20.

Ramsden, J. (2003) *The Dam Busters*, London: I.B. Taurus & Co.

Reeves, P.M. (2001) 'How individuals coping with HIV/AIDS use the Internet', *Health Education Research*, 16(6): 709–19.

Reid, A. (2005) *An Indonesian Frontier: Acehnese and other Histories of Sumatra*, Singapore: Singapore University Press.

Reid, A. (ed.) (2006) *Veranda of Violence: The Background of the Aceh Problem*, Singapore: Singapore University Press.

Renner, M. and Chafe, Z. (2007) *Beyond Disasters: Creating Opportunities for Peace*, Washington, DC: WorldWatch Institute.

Reynolds, P. (2005) 'An opportunity but no guarantee', *BBC News Online report*, 5 January online. Available: www.bbc.net.uk/2/hi/asia-pacific/4148977.stm (accessed 17 April 2007).

Rieff, D. (2003) *A Bed for the Night: Humanitarianism in Crisis*, New York: Simon & Schuster.

Roger, H. (2005) 'Why Israel fears an Iranian bomb', *RUSI Journal*, 150(1): 65–69.

Roper, T. (2004) 'The global sustainable energy island initiative', *New Academy Review*, 3(1): 147–50.

Rosato, S. (2003) 'The flawed logic of democratic peace theory', *American Political Science Review*, 97: 585–602.

Rousseau, J.J. (1756) 'Rousseau à François-Marie Arouet de Voltaire' (Lettre 424, le 18 août 1756), in R.A. Leigh (ed.) (1967) *Correspondance complète de Jean Jacques Rousseau, Tome IV 1756–1757*, Geneva: Institut et musée Voltaire, Les Délices: 37–50.

Sagar, R. (2004) 'What's in a name? India and America in the twenty-first century', *Survival*, 46(3): 115–35.

Salehyan, I. (2008) 'From climate change to conflict? No consensus yet', *Journal of Peace Research*, 45: 315–26.
Salzman, P.C. (ed.) (1982) *Contemporary Nomadic and Pastoral Peoples: Asia and the North*, Williamsburg, VA: Department of Anthropology, College of William and Mary.
Sapir, D.G. and Misson, C. (1992) 'The development of a database on disasters', *Disasters*, 16(1): 80–86.
Scanlon, J. (2006) 'Dealing with the tsunami dead: unprecedented international co-operation', *Australian Journal of Emergency Management*, 21(2): 57–61.
Schneider, S. (2005) 'Administrative breakdowns in the governmental response to Hurricane Katrina', *Public Administration Review*, 65(5): 515–16.
Schofield, V. (2003) *Kashmir in Conflict: India, Pakistan and the Unending War*, London/New York, I.B. Taurus and Company.
Schraeder, P.J. (2001) '"Forget the rhetoric and boost the geopolitics": emerging trends in the Bush Administration's policy towards Africa, 2001', *African Affairs*, 100(400, 1 July): 387–404.
Schulze, K.E. (2005) *Between Conflict and Peace: Tsunami Aid and Reconstruction in Aceh*, London: London School of Economics.
SEEDS (2007) 'Masons with a disaster risk reduction mission', in Global Network of NGOs for Disaster Risk Reduction Building Disaster (ed.) *Building Disaster Resilient Communities: Good Practices and Lessons Learned*, Geneva: UNISDR (United Nations International Strategy for Disaster Reduction): 20–22.
Shapley, D. (1974) 'Weather warfare: Pentagon concedes 7-Year Vietnam effort', *Science*, 184(4141): 1059–61.
Shawcross, W. (2000) *Deliver us from Evil: Warlords and Peacekeepers in a World of Endless Conflict*, London: Bloomsbury.
Silverstein, M.E. (1991) 'International disaster research', paper presented at International Telemedicine/Disaster Medicine Conference, National Aeronautics and Space Administration. American Institute of Aeronautics and Astronautics, Bethesda, MD, 9–11 December.
Singer, M.R. (1992) 'Sri Lanka's Tamil–Sinhalese ethnic conflict: alternative solutions', *Asian Survey*, 32(8): 712–22.
SIPRI (2009) *SIPRI Yearbook 2009*, Stockholm: Stockholm International Peace Research Institute.
Skidmore, D. (2005) 'Understanding the unilateralist turn in U.S. foreign policy', *Foreign Policy Analysis*, 2: 207–28.
Smith, A. (2003) 'A glass half full: Indonesia–U.S. relations in the age of terror', *Contemporary Southeast Asia*, 25(3): 462–63.
Smith, K. (1996) *Environmental Hazards: Assessing Risk and Reducing Disaster*, 2nd edn, London: Routledge.
Sobel, R.S. and Leeson, P.T. (2006) 'Government's response to Hurricane Katrina: a public choice analysis', *Public Choice*, 127(1–2): 55–73.
Solheim, T. and van den Bos, A. (1982) 'International disaster identification report. investigative and dental aspects', *American Journal of Forensic Medicine and Pathology*, 3(1): 63–67.
Somers, S. (2005) 'Katrina: did Federal priorities lead to a slow response?', *International Journal of Mass Emergencies and Disasters*, 23(3): 215–20.
Sommer, A. and Mosley, W.H. (1972) 'East Bengal cyclone of November, 1970: epidemiological approach to disaster assessment', *The Lancet*, 299(13 May): 1029–36.

Spence, R. (2004) 'Risk and regulation: can improved government action reduce the impacts of natural disasters?', *Building Research and Information*, 32(5): 391–402.
Sprout, H. and Sprout, M. (1957) 'Environmental factors in the study of international politics', *Journal of Conflict Resolution*, 1(4): 309–28.
Steinberg, T. (2000) *Acts of God: The Unnatural History of Natural Disaster in America*, New York: Oxford University Press.
Stokes, G.H., Evans, J.B., Viggh, H.E.M., Shelly, F.C. and Pearce, E.C. (2000) 'Lincoln Near-Earth Asteroid Program (LINEAR)', *Icarus*, 148: 21–28.
Strand, H., Wilhelmsen, L. and Gleditsch, N.P. (2004) *Armed Conflict Dataset Codebook*, Version 3.02, 20 December, Oslo: PRIO (International Peace Research Institute).
Sulistiyanto, P. (2001) 'Whither Aceh?', *Third World Quarterly*, 22(3): 437–52.
Susskind, L.E. (1994) *Environmental Diplomacy: Negotiating More Effective Global Agreements*, Oxford: Oxford University Press.
Sutton, A.E., Dohn, J., Loyd, K., Tredennick, A., Bucinia, G., Solórzanoc, A., Prihodko, L. and Hanan, N.P. (2010) 'Does warming increase the risk of civil war in Africa?', *Proceedings of the National Academy of Sciences*, 107(25): E103.
Tambiah, S.J. (1986) *Sri Lanka: Ethnic Fratricide and the Dismantling of Ddemocracy*, London: I.B. Taurus & Co.
Tanguy, J.-C., Ribière, Ch., Scarth, A. and Tjetjep, W.S. (1998) 'Victims from volcanic eruptions: a revised database', *Bulletin of Volcanology*, 60(2): 137–44.
Terry, F. (2002) *Condemned to Repeat? The Paradox of Humanitarian Action*, Ithaca, NY: Cornell University Press.
Thorner, A. (1949) 'The Kashmir conflict', *Middle East Journal*, 3(2): 164–80.
Tiranti, D. (1977) 'The un-natural disasters', *The New Internationalist*, 53: 5–6.
Toya, H. and Skidmore, M. (2007) 'Economic development and the impacts of natural disasters', *Economics Letters*, 94: 20–25.
Turcios, A.M.I. (2001) 'Central America: A region with multiple threats and high vulnerability?', *Norwegian Church Aid Occasional Paper Series*, 5.
UNDP (2004) *Reducing Disaster Risk: A Challenge for Development*, New York City: Bureau for Crisis Prevention and Recovery, UNDP (United Nations Development Programme).
UNDP (2005) *Human Development Report 2005*, New York City: UNDP (United Nations Development Programme).
UNISDR (2002) *Disaster Reduction for Sustainable Mountain Development: 2002 United Nations World Disaster Reduction Campaign*, Geneva: UNISDR (United Nations International Strategy for Disaster Reduction).
UNISDR (2004) *Living with Risk*, Geneva: UNISDR (United Nations Secretariat for the International Strategy for Disaster Risk Reduction).
UNISDR (2005) *Hyogo Framework for Action 2005–2015: Building the resilience of nations and communities to disasters*, Geneva: UNISDR (United Nations International Strategy for Disaster Reduction).
UNISDR (2009) *UNISDR Terminology on Disaster Risk Reduction (2009)*, Geneva: UNISDR (United Nations International Strategy for Disaster Reduction).
USAID (2002) *Indonesia: The Development Challenge*, Washington, DC: USAID (United States Agency for International Development).
USDS (2002) *U.S. Support for Democracy and Human Rights in Indonesia and East Timor*, Washington, DC: International Information Programs, USDS (United States Department of State).

USDS (2003) *Indonesia: Country Reports on Human Rights Practices – 2002*, Washington, DC: Bureau of Democracy, Human Rights, and Labor, Bureau of Public Affairs, USDS (United States Department of State).

USDS (2005a) *Daily Press Briefing, Sean McCormack, Spokesman, Washington, DC, 31 August 2005*, Washington, DC: USDS (United States Department of State).

USDS (2005b) *Daily Press Briefing, Sean McCormack, Spokesman, Washington, DC, 1 September 2005*, Washington, DC: USDS (United States Department of State).

USDS (2005c) *Daily Press Briefing, Sean McCormack, Spokesman, Washington, DC, 6 September 2005*, Washington, DC: USDS (United States Department of State).

USDS (2005d) *Daily Press Briefing, Sean McCormack, Spokesman, Washington, DC, 8 September 2005*, Washington, DC: USDS (United States Department of State).

USDS (2005e) *Daily Press Briefing, Sean McCormack, Spokesman, Washington, DC, 12 September 2005*, Washington, DC: USDS (United States Department of State).

USDS (2005f) *Hurricane Katrina*, Washington, DC: USDS (United States Department of State).

US Government (2002) *The National Security Strategy of the United States of America (September 2002)*, Washington, DC: The White House.

US Government (2006) *The National Security Strategy of the United States of America (March 2006)*, Washington, DC: The White House.

USGS NEIC (2005) National Earthquake Information Center – NEIC, Denver: USGS NEIC (United States Geological Survey National Earthquake Information Center): online. Available: http://earthquake.usgs.gov/regional/neic (accessed 17 October 2010).

US House of Representatives (2006) *A Failure of Initiative: Final Report of the Select Bipartisan Committee to Investigate the Preparation for and Response to Hurricane Katrina*. 109th Congress, 2nd Session, Report 109–377, Washington, DC: United States Government.

Uyangoda, J. (2005) 'Ethnic conflict, the state and the tsunami disaster in Sri Lanka', *Inter-Asia Cultural Studies*, 6(3): 341–52.

van den Bos, A. (1980) 'Mass identification: a multidisciplinary operation. The Dutch experience', *American Journal of Forensic Medicine and Pathology*, 1(3): 265–70.

van Dijk, C. (1981) *Rebellion Under the Banner of Islam: The Darul Islam in Indonesia*, Leiden: M. Nijhoff.

van Niekerk, D. (2007) 'Disaster risk reduction, disaster risk management and disaster management: academic rhetoric or practical reality?', *Disaster Management South Africa*, 4(1), 6–9.

Wachtendorf, T. (2000) 'When disasters defy borders: What we can learn from the Red River flood about transnational disasters', *Australian Journal of Emergency Management*, 15(3): 36–41.

Wagret, P. (1968) *Polderlands*, London: Methuen & Co.

Waltham, T. (1978) *Catastrophe: The Violent Earth*, New York: Crown Publishers.

Ward, R.D. (1918) 'Weather controls over the fighting during the summer of 1918', *The Scientific Monthly*, 7(4): 289–98.

Warnaar, M. (2005) 'Shaken, not stirred: Iranian foreign policy and domestic disaster', unpublished master's thesis, Universiteit van Amsterdam.

Webb, A.P. and Kench, P.S. (2010) 'The dynamic response of reef islands to sea-level rise: Evidence from multi-decadal analysis of island change in the Central Pacific', *Global and Planetary Change*, 72: 234–46.

Weeramantry, C.G. (1992) *Nauru: Environmental Damage under International Trusteeship*, Melbourne: Oxford University Press.

Weizhun, M. and Tianshu, Q. (2005) 'Disaster diplomacy: a new diplomatic approach? The apocalypse of the Indian Ocean earthquake and tsunami', *World Politics and Economy* (Chinese Academy of Social Sciences), 6: 111–24.

Wheeler, D. (1991) 'The influence of the weather during the Camperdown campaign of 1797', *The Mariner's Mirror*, 77(1): 47–54.

Wheeler, D. (1995) 'A climatic reconstruction of the Battle of Quiberon Bay, 20 November 1759', *Weather*, 50(7): 230–39.

Wheeler, D. (2001) 'The weather of the European Atlantic seaboard during October 1805: an exercise in historical climatology', *Climatic Change*, 48: 361–85.

Wheelis, M. (2002) 'Biological warfare at the 1346 siege of Caffa', *Emerging Infectious Diseases*. 8(9): 971–75.

WHO (2003) *The World Health Report 2003*, Geneva: World Health Organization.

Williams, C.C. (1996) 'Local purchasing schemes and rural development: an evaluation of local exchange and trading systems (LETS)', *Journal of Rural Studies*, 12(3): 231–44.

Wilson, A.J. (2000) *Sri Lankan Tamil Nationalism: Its Origins and Development in the 19th and 20th Centuries*, London: C. Hurst and Co.

Wisner, B., O'Keefe, P. and Westgate, K. (1977) 'Global systems and local disasters: the untapped power of people's science', *Disasters*, 1(1): 47–57.

Wisner, B. (2003) 'Sustainable suffering? Reflections on development and disaster vulnerability in the post-Johannesburg world', *Regional Development Dialogue*, 24(1): 135–48.

Wisner, B., Blaikie., P., Cannon, T. and Davis, I. (2004) *At Risk: Natural Hazards, People's Vulnerability and Disasters*, 2nd edn, London: Routledge.

Yalçinkaya, A. (2003) 'From disaster solidarity to interest solidarity: Turkish–Greek relations after the Marmara and Athens earthquakes within the concept of game theory', *Turkish Review of Balkan Studies*, 2003: 149–202.

Yim, E.S., Callaway, D.W., Fares, S. and Ciottone, G.R. (2009) 'Disaster Diplomacy: Current Controversies and Future Prospects', *Prehospital and Disaster Medicine*, 24(4): 291–93.

Yin, R. and Baek, J. (2004) 'The US–Canada softwood lumber trade dispute: what we know and what we need to know', *Forest Policy and Economics*, 6(2): 129–43.

Zaman, M.S. (1991) 'Silent disasters of Africa', *World Health*, Jan.–Feb.: 7–9.

Index

Abramovitz, J. 12
Aceh 13, 16, 40–7, 71; further studies 145; limitations 134; qualitative typologies 83, 85, 88, 93, 99
ACNDR 136
active use of diplomacy 97, 100, 127
active use of weather 93–7
Aegean Sea 82
Afghanistan 23, 37–8, 49, 64; failures 114, 117; qualitative typologies 91
Africa 6, 12, 26–7, 48; case studies 55; limitations 134; quantitative analyses 71
aggregation 78–9, 81
agriculture 27, 72
aid relationship 85–7, 97, 100
Akcinaroglu, S. 115–16, 147
Alaska 55
Alexander the Great 5
Algeria 24, 91
AmeriCare 21
analyses 69–101
Andaman Islands 45
Anguilla 123
Antarctic Treaty System (ATS) 81, 119–20
Antarctica 120
anti-diplomacy 126
anticipation 99
apartheid 26
appeals 79
application of diplomacy 135–41
Aral Sea 55
Armed Conflict Dataset 1, 73
Armitage, R. 22–3, 90
Asians 61
Aspinall, E. 42
astronomical objects 63, 149
Atlantic 31, 85

Aung San Suu Kyi 53
Australia 24, 40, 48, 57–8; case studies 61–2; qualitative typologies 91; spin-offs 123–4
Autesserre, S. 116
Avian flu 63
avoid forcing 103–4
avoiding diplomacy 110–11, 120
Axis of Evil 21, 28, 92, 112

Balamir, M. 2
Bali 48, 62
Balkan Pact 32
Baltic Sea 62
Ban Ki-moon 53
Bangladesh 5–6, 60, 74, 106, 112, 116
banks 58
bathymetry 70
Belgium 74
bereavement 61
Berger, M.T. 42
bias 133–4, 139
biodiversity 121
biological warfare 94
biosphere 122
Bird, D.K. 133
BJP 37
Black Death 94
Bolivia 84
bombs 39, 41, 46, 48, 62, 94, 117
Bonaparte, N. 93
border wars 34, 36, 38, 56
boreholes 94
Bose, S. 38
Box, G.E.P. 76
Brancati, D. 69–72
Broder, S. 5
Bronze Age 5
Brown, M. 50

Brunei 58, 121
Buhaug, H. 69, 72–3
building blocks 79, 82
building codes 12, 70
bureaucracy 43
Burke, M.B. 69, 71–3
Burma 45, 52–5, 148
Bush, G.W. 21–3, 28, 46–9, 63, 91, 112

5C+A framework 98–9
Caffa 94
Cambodia 95
Canada 13, 40, 58, 78, 121–3
Canary Islands 60
capabilities 98
capital punishment 57
Caribbean 5, 84, 123
Carroll, M.S. 115
Carteret Islands 55
case studies 3, 5–7, 9–10, 12–13;
    empirical evidence 18–68; failures
    116; further 144–5; future trends 147;
    implementation 141; lessons 136–7;
    levels 139–40; limitations 128, 130–2,
    142; organising 18–20; qualitative
    typologies 77–89, 92–3, 95–6, 98–101;
    quantitative analyses 70; questions
    15–16; spin-offs 123–5; successes
    104–5; summing up 66–8; tit-for-tat
    107–8
Caspian Sea 21, 55
Castro, F. 15, 30–1, 80, 89–90, 138, 147,
    149
casualty identification 18–19, 60–3
catalyses 13–16, 67, 117, 127
Cater, N. 21
Caucasians 61
Cayman Islands 123–4
Center for Humanitarian Dialogue 145
Central America 118
Chafe, Z. 43
Charles XII of Sweden 93
checklists 15–16
checkpoints 39
chemical weapons 20
China 5, 24, 27, 29; case studies 38, 40,
    54, 58; failures 116; lessons 138;
    mirror diplomacy 108; qualitative
    typologies 86, 91; spin-offs 121;
    tit-for-tat 107
cholera 118
Christianity 65
Civil Contingencies Bill 136
civil wars 5, 71–3, 83, 116

Clifford, R.A. 4
climate affairs template 138
climate change 19, 30, 55, 60; failures
    117; quantitative analyses 72, 76;
    spin-offs 122
Clinton, B. 107
cloud seeding 94–5
Cold War 28–30, 63–4, 72, 118
Collaborating Centre for Research on
    the Epidemiology of Disasters
    (CRED) 74
collisions 28, 60–2
Colombia 84
combined aid 85–7
comets 63
Comfort, L. 6, 9, 78–9
Commonwealth 123, 126
communism 20, 26, 85
community cohesion 4–5
compassion 98
competence 98
complex adaptive systems 78–82, 100
conflict likelihood 69–72
conflict resolution 8, 13–14, 36, 43–4;
    case studies 48; qualitative typologies
    88; spin-offs 121; vaccination
    programmes 63–4
conflict zones 1, 12, 83–5, 94;
    exacerbation 114–18; failures 115–18;
    further studies 144; future trends
    148–9; qualitative typologies 79;
    spin-offs 121–2; successes 104
confounding factors 131–3
Congo River 83, 139
Conrich, B. 125
Convention on Biological Diversity 81
Convention on the Conservation of
    Antarctic Marine Living Resources
    120
Convention for the Conservation of
    Antarctic Seals 119
Cook Islands 123
correctness 98
corruption 25, 47, 54, 124, 139
Corruption Perceptions Index 139
Cotton, W.R. 94
credibility 99, 105
cricket 38, 40
criminal codes 57
crisis management 109
cross-border disasters 4, 13
Cuba 6, 12, 15, 30–2; case studies 48,
    50, 52, 66; failures 110–13, 116; future
    trends 147; implementation 141;

lessons 137–8; levels 140; limitations 129; mirror diplomacy 108; qualitative typologies 80–7, 89–90, 95; successes 102, 105
currency 58
cycle model 2–3
cyclones 5, 18, 30, 45; case studies 52–4; failures 112; future trends 148; qualitative typologies 94–5; quantitative analyses 74
Cyprus 32, 34, 125–6, 129

dams 28, 76, 94
databases 74–6
Davidson, W.D. 88
De Boer, J.Z. 5
death penalty 57
deep sea 120
deforestation 55
democracy 8, 20, 45, 49; application 137; case studies 53–4; quantitative analyses 76; role 142
Democratic Republic of Congo (DRC) 64, 83–4
democratization 72
Democrats 114
dependency on disaster 111
Der Derian, J. 126
desertification 55
détente 23, 90
Diamond, L. 88
Diaz, N. 108
dictatorships 20, 45, 53, 89, 117
diplomacy, definitions 3, 13, 119
disaster, definitions 1–2, 13, 119
disaster diplomacy 1–6; analyses/typologies 69–101; application 135–41; development 7–10; failures 110–22; further studies 144–6; future trends 147–9; history 4–6; levels 139–41; limitations 127–34, 142–4; purpose 89–92, 97, 100, 126; successes 102–10, 119–22, 146
disease 5, 29, 63–6, 75
displaced people 73
dissertations 8
distractions 120
diversity 78, 131
Dole, E. 21
Dominican Republic 84
donor-recipient aid 86–7, 105
Dove, M.R. 6
dracunculiasis 63–5
Draper, N.R. 76

drought 5–6, 12, 26–8, 30; case studies 34–6; failures 111; limitations 129, 132, 134; mirror diplomacy 108; qualitative typologies 78, 82, 86, 95; quantitative analyses 70, 73–4
drug trafficking 38, 40
Drury, A.C. 5–6
Dubai 59
Dutch 42, 61, 93
Dynes, R.R. 4

earthquakes 1, 5–6, 11, 18; case studies 21–4, 32–4, 36, 38–40, 46, 54, 60; casualty identification 62; failures 112–13, 115–16, 118; future trends 148; lessons 138; levels 139–40; limitations 129, 134; qualitative typologies 78, 81–2, 86–8, 90–2, 94; quantitative analyses 69–71, 73–4; successes 106; tit-for-tat 106–8
East Pakistan 112, 116
East Timor 107–8, 124
Eastern Bloc 20
economics 27, 42, 47, 54; Africa 72; case studies 58, 64; instability 76; lessons 138; qualitative typologies 78, 81; spin-offs 120
ecosystems 121–2
Ecuador 118, 138
education 2, 16, 39, 88
Egypt 21–3, 36, 62
email 79
embargos 30, 110
Emergency Events Database (EM-DAT) 74–6
emergency management 31, 41, 50, 70, 114, 136–7
empirical evidence 18–68
England 64
English school 2, 8
Enia, J. 43
Enniskillen 58
environmental management 25; change 55–6, 59–60; conflict theory 8; diplomacy 119–22; limitations 131; quantitative analyses 76–7; spin-offs 119, 122; warfare 95
Environmental Modification Convention 95
epidemics 75, 94, 118, 145
Eritrea 34–6, 42, 56, 82; failures 113, 116; future trends 149; implementation 141; limitations

128–9; qualitative typologies 88, 96, 99
erosion 56
ethics 10, 54, 97, 109, 127–31, 138
Ethiopia 34–6, 42, 56, 82; failures 113, 116; future trends 149; implementation 141; limitations 128–9; qualitative typologies 88, 96, 99
euro 33
Europe 32, 61, 93–4
European Union (EU) 33, 40, 44, 82, 86
Europeanisation 33, 92
exacerbation of conflict 114–18
exercises 137
expectations 111

failure pathways 110–18
false propinquity 111, 113, 125
famine 28–9, 36, 70, 73, 83, 108
Federal Emergency Management Agency (FEMA) 50
ferries 62
field work 18
Fiji 58, 60
floods 1, 4, 13, 25; case studies 27–8, 41, 56; failures 115; lessons 139; limitations 132; mirror diplomacy 108; qualitative typologies 84, 86, 93–5, 101; quantitative analyses 73–4
Florida 31, 89–90, 112
flow 78–9, 81–2
focus 103–4, 120
food market 27–8, 64
forensic teams 61
Fortes, M. 121
fossil fuels 121–2, 125
Fouta Djalon 55
fragmentation 76
framework of action 102–3
France 53, 64
fraud 48
French language 84
fundamentalism 20
further failures 114–18

Gaillard, J.-C. 16, 24, 43–4
Gambia River 55
Ganapti, E. 33, 92, 140
Gandhi, R. 41
GAO 136
Gawronski, V.T. 4, 98–9
GDP 55, 69, 73
General Electric Corporation 95

genocide 83
geography 18–19, 26, 66, 83–4; qualitative typologies 87
geology 73–4
Gerakan Aceh Merdeka (GAM) 42–3
Germany 94
Glantz, M.H. 5–6, 9, 30, 32, 66, 78, 81, 94–5, 105, 129, 137–8, 147
global disasters 63
Goma 83
González, E. 15, 112
government-led disaster diplomacy 88–9, 102
Great Barrier Reef 57
Greece 5–6, 32–4, 60, 78; failures 113; future trends 148–9; lessons 138; levels 140; limitations 128–9; qualitative typologies 80, 82–3, 87–8, 92–3; successes 102, 105; tit-for-tat 107–8
greenhouse gases 55, 76
groundwater 94
Guatemala 5
guerrillas 25–6, 85
guidelines 137
Guinea 55, 107
Gulf Coast 13, 85
Gulf of Oman 20
Gusmao, X. 107

Haiti 84, 118
Hannibal 94
Hartmann, B. 60
health 29, 63, 65–6, 75
Health as Bridge for Peace programme 63
hegemony 48–9, 52
Henderson, D.A. 65
hidden victims 76
Hispaniola 84
history 3–6, 16, 19, 40; case studies 50; environmental 55; future trends 147; limitations 130–1; qualitative typologies 81–2, 84, 89, 94; quantitative analyses 74; spin-offs 123; vaccination programmes 63
Hitler, A. 93
HIV/AIDS 65–6, 75
Hoffman, S. 75
Holland, J.H. 78
Holloway, A. 6, 9, 26–7, 78
homosexuality 65–6
Honduras 5, 115
Hotez, P.J. 29, 64–5

Human Development Index 107
human rights 23, 35, 47, 53–4; case studies 57–8; limitations 130–1; qualitative typologies 91
Hurricane Katrina 13, 21, 24, 28; case studies 31, 48–52, 60, 62, 66; future trends 148; lessons 136–7; mirror diplomacy 109; qualitative typologies 86–7; tit-for-tat 106–7
hurricanes 12, 31, 74, 81; failures 115–16, 118; lessons 137; limitations 129, 131; mirror diplomacy 108; qualitative typologies 85, 87, 90, 95; spin-offs 123
Hussein, S. 20–1, 117
hypotheses 11–17

India 5, 36–40, 45–6, 58; case studies 60; failures 111–13; further studies 145; future trends 148; limitations 132, 134; qualitative typologies 83, 86, 88, 93; successes 102, 106; tit-for-tat 108
Indian Ocean 40, 46, 48, 88, 106
indigenous people 58, 84, 120
Indonesia 40, 42–3, 46–8, 58; case studies 61–2; further studies 145; qualitative typologies 93; spin-offs 124
informal networks 103–5, 122–3
internal model 79, 82
International Center for Desert Affairs 138
international community 38, 40, 49, 56
international relations 2–3, 8, 79, 119–20, 144
internet 54, 81
inverse disaster diplomacy 113–14
invisible disasters 75–6
Iran 5, 18, 20–4, 47; case studies 50, 52, 60; failures 112–13; further studies 145; future trends 148; levels 139–40; limitations 129, 132; qualitative typologies 83, 86–7, 90–2, 96; tit-for-tat 106–7
Iraq 20–1, 49, 86, 91, 114, 117
Islam 65, 85
island communities 55–60, 66
isolationism 48–50, 52, 58, 81, 109
Israel 12, 21–3, 36, 62–3, 90–1

Japan 24, 27–8, 58, 61–2; failures 115; mirror diplomacy 108; qualitative typologies 83, 85–6, 91; spin-offs 121; tit-for-tat 107
Java 43

Jenner, E. 64
Jordan 12, 22–3, 63
journals 6

Kahan, A.M. 94
Kalla, Y. 43
Kardol, R. 59
Karp, E. 62
Kashmir 38–40, 111, 134; future trends 148
Kelman, I. 6–7, 9, 11–12, 15, 23, 45, 50, 63, 79, 81–2, 88, 90, 96, 123, 125, 135
Kenedi, A. 15
Ker-Lindsay, J. 6, 9, 32–3, 78, 92, 129, 140
Kermadec Islands 57
Khan, M.H. 6
Khatami, M. 22
Khomeini, R. 20
Kim Dae-jung 28
Kim Jong-il 28
Kiribati 56–7
Korean War 27, 32
Kosovo 32, 82
Kouchner, B. 53
Koukis, T. 6–7, 9, 11–12
Kuala Lumpur 122
Kuril Islands 121
Kuwait 20, 117

landslides 74, 131, 145
Laos 95
Latin America 138
Layang Layang 59
Le Billon, P. 43, 115
Lebanon 20
lessons for application 135–41
levels 103–5, 123
levels of diplomacy 139–41
Lewis, J. 2, 5, 75, 93, 112, 114
Liberation Tigers of Tamil Eelam (LTTE) 41–2
Liberia 64
limitations 127–34, 142–4
Line of Control 38–9
literature 3–4, 7–9, 18, 27; case studies 33, 60; casualty identification 60; failures 115–16; further studies 144, 146; future trends 147–8; limitations 130, 142; qualitative typologies 78, 88; quantitative analyses 71–3, 76; spin-offs 125; vaccination programmes 64, 66

litigation 95
livelihood concerns 16–17, 72, 81, 84, 121
lobbying 96, 123, 148
Local Exchange Trading Systems 58
log books 93
logging 25
logistics 35, 62, 120

McDonald, J. 88
Macquarie Island 70
malaria 75
Malaysia 58–9, 121–2
Maldives 45, 56, 58, 106
Mandel, R. 115
Mardhatillah, F. 42
Martin, B. 105
Martinique 5
Mavrogenis, S. 33, 92
media 3, 6, 8, 18; case studies 21–2, 25, 32–3, 37, 39, 43, 50, 54–5; further studies 145; future trends 148–9; lessons 138; limitations 128; qualitative typologies 78, 81–2, 88, 92, 96; quantitative analyses 75; successes 102, 105
meteorites 12, 63, 74
meteorology 73–4, 79, 138
Mexico 84
micro-diplomacy 125
Middle East Regional Cooperation Program 12, 137
migrants 56–9, 76, 121
Miller, J. 15
mirror disaster diplomacy 102, 108–9
mobile islands 59
Montevideo convention 59
Montserrat 115, 123–4
Montville, J.V. 88
Morfit, M. 43
morphology 70
motivation 146
Mozambique 26, 115
Mubarak, H. 21
multi-track diplomacy 89, 103–5, 123
multi-way process 103, 105, 123
multiculturalism 138
multinational corporations 64
Musharraf, P. 37
Muslims 25, 40
mutual aid 85
Myanmar 52
mytho-diplomacy 126
myths 5, 105

Namibia 26
NATO 32, 37, 49, 86, 117
Nauru 124
Nel, P. 5, 69, 73–4
Nelson, T. 91, 110–12, 115
neo-diplomacy 126
Netherlands 62, 94
New People's Army (NPA) 25–6
New Zealand 40, 57–8, 62, 70, 136
news 75, 81, 113
Nicaragua 5
Nicobar Islands 45
Niger River 55
El Niño-Southern Oscillation (ENSO) 30
Niue 123
Nobel Peace Prize 28
nomads 59
non-linearity 78, 80–1
non-sovereign jurisdictions 122–6
North Korea 18, 27–9, 47, 64; case studies 66; failures 112; lessons 138; mirror diplomacy 108–9; qualitative typologies 81, 83–6; spin-offs 120; tit-for-tat 107
North Sea Campaign 93
Northern Cyprus 125
Northern Ireland 58
nuclear energy 22–3, 46, 91
nuclear weapons 20–1, 28, 37–8, 46, 94, 107

Obama, B. 133
oil 20, 47, 58, 120–1
O'Keefe, P. 11
Olson, R.S. 4–6, 98–9
Olympic Games 54–5
organisation-led disaster diplomacy 88–9, 102, 104–5
origins of disaster diplomacy 1–6
Ottoman Empire 32
outcomes 24, 26, 36, 52; case studies 63; limitations 127–9, 131; qualitative typologies 78, 89, 99; successes 104; vaccination programmes 65–6
outer space 120
overwhelming events 111

Pacific 48, 57–8, 123–4
Pakistan 5, 36–40, 46, 60; case studies 62; failures 111–13; further studies 145; future trends 148; limitations 134; qualitative typologies 83, 88; successes 102; tit-for-tat 108

Palestinian National Authority 12
Palestinians 12, 63
Palm Islands 59
pandemics 1, 63
Papua New Guinea 55
para-diplomacy 119, 122–6
passive use of diplomacy 97, 100, 127
passive use of weather 93–7
Peace Parks 121–2
Pelée, Mt 5
people-led disaster diplomacy 88–9, 102
Persian Gulf 20
Peru 5, 118, 138
Petropoulos, N.P. 32, 105, 144
Philippines 5, 13, 16, 18; case studies 24–6, 58; qualitative typologies 83, 85; spin-offs 121
Pielke, R.S. Sr. 94
Pinatubo, Mt 16, 24
Pitcairn Island 81
planes 60–2
Poland 115
police 61–2, 84
polio 63–5
political science 8, 36
politics 3, 5–6, 8, 14–17; application 135; case studies 21–7, 30–1, 37–8, 40–1, 43–8; casualty identification 61, 63; failures 113–14, 118; further studies 145–6; instability 76; lessons 137–9; levels 140; limitations 127–31, 134, 143; mirror diplomacy 108–9; qualitative typologies 78–81, 83–4, 86, 88, 90–2, 96; quantitative analyses 72; spin-offs 119, 125; successes 102, 105–6; tit-for-tat 107; vaccination programmes 65–6
population density 11, 69, 77
portfolios 103
Portugal 11
poverty 64, 76
Powell, C. 21–2, 47, 52, 113
precipitation 71, 95
predictive-model issues 98–100, 132
Premadasa, R. 41
Prince Edward Island 123
profit 59, 124
propaganda 28–9, 35
propinquity 83–5, 97, 100, 111, 113, 120, 125
proto-diplomacy 125–6
Protocol on Environmental Protection 119
protocols 61–2, 119

proxy wars 72
public opinion 47, 88, 92, 105
public relations 55
Puerto Rico 60

al-Qaeda 117
Qian Ye 138
qualitative typologies 9, 68–9, 77–98, 142
quantitative analyses 5, 9, 68–77, 99–100, 142
Quarantelli, E.L. 4
Quiberon Bay, battle of 93

rainfall 27, 35, 72, 74, 95
Rajapaksa, M. 42
Reagan, R. 65
reciprocity 33, 78–9, 106–7
La Red 2, 75
Red Crescent 92, 143
Red Cross 21, 29, 92, 143
refugees 60, 83
Reid, A. 42
relations 111
religion 20, 39, 42, 50, 72, 88
remittances 81
Renner, M. 43
repatriation 60
repertoire 103
Republic of Ireland 58
Republicans 21, 50, 114
research questions 11–17
reservoirs 94
resource scarcity 71–2, 76, 95
rhetoric 37
Richter scale 69–70
Righarts, M. 5, 69, 73–4
Rio Grande River 4
risk reduction 2, 12, 25, 31; case studies 44, 54; further studies 145–6; future trends 149; implementation 141; lessons 137, 139; limitations 132, 143; mirror diplomacy 108–9; qualitative typologies 84, 99; spin-offs 124, 126; successes 106
Robertson, P. 50
Ronfeldt, D. 15
Rousseau, J.J. 11
Russia 84, 93–5, 107–8, 121
Rwanda 83

Sahel 5
St Helena 123
salinisation 55

Sanders, D.T. 5
Sandinistas 5, 116
Saskatchewan 122
Saudi Arabia 20
Scanlon, J. 61–2
schadenfreude 105
Schofield, V. 38
Schulze, K.E. 43
science 103, 105, 122–3, 140
Scotland 58
sea levels 19, 55–60, 66, 72, 76
search-and-rescue teams 22, 54, 81
seismologists 69
Senegal River 55
September 11 2001 28, 30, 37–8, 45–7, 49–50, 62, 117, 132
Serbia 32
sexuality 66
Shah of Iran 20
shipping lanes 121–2
ships 93
Sierra Leone 64
silent disasters 75–6
Silverstein, M.E. 6
Singapore 61
Sinhalese 40–1
Skye 58
slavery 5
Small Island Developing States 84
smallpox 63–5
smuggling 121
social construction of disasters 11–12
sociology 4, 144
Solomon Islands 58
Somalia 45
South Africa 26
South America 118
South China Sea 57, 121
South Korea 28–9, 62, 81, 85; mirror diplomacy 108; qualitative typologies 86; tit-for-tat 107
southeast Asia 47
southern Africa 26–7, 55, 78, 134
sovereignty 54, 56–7, 59, 124–6
Spain 61
Spanish language 84
Sparta 5, 116
spin-offs 119–26, 149
spotlights 111, 113, 129
Spratly Islands 57–9, 121–2
Sri Lanka 40–6, 58, 61–2, 70–1; qualitative typologies 78, 83, 85, 93, 99; successes 104, 106; tit-for-tat 106

state-involvement types 92–3, 97, 100
states of emergency 74–5
Stevenson, R.L. 93
stockpiling 70
storm surges 27–8, 73–4
sub-Saharan Africa 71–2
success pathways 102–9
Sudan 64, 83, 85, 116–17
suicide attacks 39, 41, 62
Sulistiyanto, P. 42
summaries 97–8, 100–1
summits 37, 107
superpowers 72
sustainability 8, 36, 44, 102, 131
Sutton, A.E. 72
Sweden 40, 62, 93
swine flu 29
symbolism 103, 105–6, 122–3

tagging 79, 82
Taiwan 54, 58, 116, 121
Taleban 38, 117
Tamils 40–2, 85, 104
Tanguy, J.-C. 73
Tanzania 123
Tasmania 123
techno-diplomacy 126
temperature projections 72
Tenerife 60–2
terrorism 21, 28, 30, 37–9; case studies 41, 46, 48–50, 62; failures 117; lessons 137; limitations 132; successes 109
texting 79
Thailand 61
Thera 5
Thompson, T.G. 64
Tiranti, D. 11
tit-for-tat disaster diplomacy 102, 106–8
Togo 107
toolkit 103
topography 70
tornadoes 81, 113–14
tourism 45, 48, 57–9, 62, 81
tracks for diplomacy 88–9, 97, 100, 103–5, 123, 137–9
trade 31, 45–7, 79, 110, 122–3, 138
training 70, 88, 137–8
trains 28, 39
Transparency International 139
treason 45
triangulation 142
Tristan da Cunha 81, 123–4

174   Index

tsunamis 11–12, 16, 38, 40–8; case studies 52, 61; further studies 145; future trends 148; limitations 134; qualitative typologies 78, 83, 86, 88, 93, 99; quantitative analyses 70–1, 74; successes 104, 106
Turcios, A.M.I. 12
Turkey 5–6, 32–4, 60, 78; failures 113, 115; future trends 148–9; lessons 138; levels 140; limitations 128–9; qualitative typologies 80, 82–3, 87–8, 92–3; spin-offs 12; successes 102, 105; tit-for-tat 107–8
Turks and Caicos Islands 122
Tuvalu 56
tweeting 79
typhoons 25–6, 74, 93
typologies 69–101

UCDP/PRIO Armed Conflict Dataset 1, 73
Uganda 83
Ukraine 94
UNESCO 21
unilateralism 49–50, 52
UNISDR 2, 12, 146
United Arab Emirates (UAE) 24, 91
United Kingdom (UK) 21, 40, 62, 86; failures 117; lessons 136; qualitative typologies 94; spin-offs 122–4
United Nations (UN) 8, 20–4, 29, 37; case studies 49, 53; failures 117; qualitative typologies 79, 81, 88, 90–1; spin-offs 123; tit-for-tat 107; vaccination programmes 63–4
United States Agency for International Development (USAID) 12, 44, 47
United States (US) 4, 6, 12–13, 15–16; Army 95; case studies 18, 20–5, 28–32, 37, 40, 45–8, 53, 66; casualty identification 62; Congress 23; Defense Department 95, 109; failures 110–13, 115–17; further studies 145; future trends 147–8; hurricanes 48–52; implementation 141; lessons 136–8; levels 139–40; limitations 129, 132; migrants 58; mirror diplomacy 108–9; National Guard 114; qualitative typologies 80–7, 89–93, 96; Senate 22; spin-offs 121, 124; successes 102, 105; tit-for-tat 106–7; vaccination programmes 63–6
USDS 47, 50, 52, 107
USSR 20, 30, 64

vaccination programmes 18–19, 29, 63–6, 116, 134, 145
Vajpayee, A.B. 37
van den Bos, A. 61
van Dijk, C. 42
van Niekerk, D. 2
Vanuatu 58
Venezuela 31, 50, 52, 137
Vietnam 58, 95, 121
vindictiveness 111, 113
visas 132
Viti Levu 60
volcanic eruptions 5, 16, 24–5, 63; failures 115; limitations 131; qualitative typologies 81, 83–4; quantitative analyses 73–4; spin-offs 123; successes 104
Voltaire 11

Waizenegger, A. 43, 115
war crimes 35
Warakaulle, C. 6
Ward, R.D. 93
warmongers 71
Warnaar, M. 23, 82, 90–1, 129, 133
water 55, 59–60, 76, 84; lessons 138; limitations 132; qualitative typologies 94–5
weapons 20–1, 111, 122, 130
weather modification 93–7
websites 6, 75, 81
Weeramantry, C.G. 124
West 20–1, 48, 117
Wheeler, D. 93
White House 22–3, 29, 48–9, 52, 114
wildfires 115
windstorms 74
Wisner, B. 2
World Health Organization (WHO) 63, 74
World Heritage sites 21
World Trade Center 30, 62, 117
World War I 32, 93
World War II 5, 20, 32, 42, 94
worsening relations 111, 122

Yemen 107
Yim, E.S. 142
youth 77
Yudhoyono, S.B. 43

Zaire 83
Zanzibar 123
Zarif, J. 22, 24, 91

WITHDRAWN

College Lane, Hatfield, Herts. AL10 9AB
Information Hertfordshire
Services and Solutions for the University

For renewal of Standard and One Week Loans,
please visit the web site **http://www.voyager.herts.ac.uk**

This item must be returned or the loan renewed by the due date.
A fine will be charged for the late return of items.